Developments in Chemogenomics

Developments in Chemogenomics

Dale H. Trimm

Oxford Book Company
Jaipur ▪ New Delhi

ISBN : 9789355247278

First published in India, 2025

Oxford Book Company
Regd. Off. "Bony Residency", Gate No. 2
Opp. Tilak Public School, Vishweswariya Nagar,
Gopalpura Road, Jaipur - 302018, India
Ph: 0141 - 2761280, Telefax: 2761381
email - oxfordbookjpr@gmail.com
website - oxfordbookcompany.com

Delhi Office: 102, 1st Floor, Satyam House,
4327/3, Ansari Road, Darya Ganj,
New Delhi - 110002, India
Ph: 011-45652440

Distributed By
KEYA INTERNATIONAL PRESS LLP
E-mail: keyainternationalpressllp@gmail.com

Printed and bound in India by Thomson Press Ltd., New Delhi

Contents

Preface *vii*

1. **Methods of Chemogenomics** **1**

Current and Future Research • Relation to Other Fields • Criticisms • Theories • Methodologies • Issues and Areas of Research • The Goal • Mechanism • Major Steps • Measuring and Detecting Transcription • Processing • Nuclear Polyadenylation • Splicing athways • Alternative Splicing Mechanisms • Alternative Splicing and Disease • Simple Illustration of an Intron and RNA Splicing

2. **Overarching Structure** **67**

Computational Methods for Studying Protein Folding • Secondary Structure • Algorithm • Tertiary Structure • Quaternary Structure • Relationship to Docking Methods • Selection of Target Proteins • Design • Structural DNA Nanotechnology • Functional Nucleic acid Nanostructures • Materials and Methods • History of DNA Research • DNA Error • Postulates

3. **Important Cellular Activities** **107**

Analytical Technologies • Experimental Techniques • Imaging Techniques • Relation to Proteome • Tools and Methods • Phosphoproteomics in the Study of ignal Transduction • Protein Databases • Organization of UniProt Databases • GenBank • Structures Outside the Cell Wall • The Basic Microarray • Microarrays and Bioinformatics • Chain-termination Methods • Amplification and clonal selection • Technology • Applications

4. **Methods Based on Knowledge** **169**

Regulated Stages of Gene Expression • Modelling • The Boolean Logic Assumption • Identification and Computational Prediction of CRMs • Tools and Features

• Predicting the Number and Organization of Operons • Associated disciplines • Identification and Characterization • Enhancers and the Evolution of Developmental Mechanisms • Molecular Level Simulations • Application • Procedure • Single Gene Disorder • Background • Nomenclature • Relationship To Other Genetic Concepts • Epistasis and Sex

5. Elements of Chemogenomics **245**

Medical Procedure • Risks and Limitations • Ethics and Law • Types and Origin of Pseudogenes • Microarray Technique • GEO • Conclusion • Dominant and Recessive Genetic Diseases in Humans • Genes • Mutations • Use in Identification • Linkage Overview • Sex-linked Traits in Other aAnimals • Traits Previously Believed to be Mendelian • X-linked Hypophosphatemia • Cause and Genetics • Sex Differences In Phenotype/Genotypes and Frequency • Y linkage • Symptoms • Sources and Storage of Cells • Conditioning Regimens • Engraftment • Complications • Presentation and Diagnosis • Diagnosis • Treatment and Prevention • Investigational Therapies for Graft-versus-host Disease • Risks to Donor • Donor Registration and Recruitment • Experimental HIV treatment • Potential Treatments

Bibliography 309

Index 311

Preface

The process of systematically screening targeted chemical libraries of small compounds against certain drug target families (such as GPCRs, nuclear receptors, kinases, proteases, etc.) with the ultimate goal of discovering new medications and drug targets is known as chemogenomics.Certain components of a target library are usually well characterised, meaning that their functions have been established and that chemicals that alter those functions—such as ligands for receptors, enzyme inhibitors, or ion channel blockers—have been found. Other target family members are categorised as orphan receptors because they may have an unidentified function and no known ligand. The role of these novel targets can be clarified by finding screening hits that influence the activity of the less well characterised members of the target family. Moreover, drug discovery can be initiated using the findings for these targets. Include known ligands of at least one, and preferably multiple, members of the target family when building a targeted chemical library. The compounds in a targeted chemical library should collectively bind to a large percentage of the target family since some ligands that were created and synthesised to bind to one family member will also bind to other family members. The study of proteins and signal transduction pathways through the screening of chemical libraries containing small compounds is known as chemical genetics. Additionally, unique chemical functional groups and tags can be introduced into proteins selectively using amino acids containing boronate, keto, acetylene, azide, thioester, and dehydroalanine. Activity-based reagents have been created to label the enzymatically active form of proteins, allowing for the investigation of enzymatic activity as opposed to total protein (see Activity-based proteomics). Serine hydrolase and cysteine protease inhibitors, for instance, have been transformed into suicide inhibitors.By direct targeting, this approach improves the capacity to analyse low abundance elements selectively. Modifications such as an identification handle or mass tag that will facilitate proteomic analysis could be added to structures that resemble these inhibitors.Converted substrate can also be used to track enzyme activity.This tactic depends on the use of artificial substrate conjugates with moieties that particular enzymes can react with.

After that, an affinity reagent is used to capture and analyse the product conjugates. The enzyme velocity can be calculated using the product conjugate concentration measurement. Understanding the signal transduction pathways in cells depends on the identification of enzyme substrates, of which there may be hundreds or thousands, many of which are unknown. Chemical biologists can address this issue by developing methods for labelling the cellular substrates of enzymes. An artificial ATP analogue is used by "analog-sensitive" kinases in a proposed way to label substrates, making it easier to visualise and identify them with a distinct handle.

This book investigates novel concepts in this field.

—Dale H. Trimm

1

Methods of Chemogenomics

Current and Future Research

Since one of the goals of DIT is to explain the distribution of human cultural traits, ethnographic and ethnologic techniques may also be useful for testing hypothesis stemming from DIT. Although findings from traditional ethnologic studies have been used to buttress DIT arguments, thus far there have been little ethnographic fieldwork designed to explicitly test these hypotheses. A major difficulty in using existing ethnographic data to test DIT is that traditional ethnographic data tends to ignore individual and intragroup cultural variation and focus almost entirely on intergroup variation. DIT has been viewed as having great potential for unifying diverse academic fields under one overarching theory. Mesoudi, et al. have identified DIT as the ideal way to build a comprehensive theory of cultural evolution to answer questions about human behavior at different temporal and spacial scales.

Along with game theory, Herb Gintis has named DIT one of the two major conceptual theories with potential for unifying the behavioral sciences, including economics, biology, anthropology, sociology, psychology and political science. Because it addresses both the genetic and cultural components of human inheritance, Gintis sees DIT models as providing the best explanations for the ultimate cause of human behavior and the best paradigm for integrating those disciplines with evolutionary theory. In a review of competing evolutionary perspectives on human behavior, Laland and Brown see DIT as the best candidate for uniting the other evolutionary perspectives under one theoretical umbrella.

Relation to Other Fields

Sociology and Cultural Anthropology

Two major topics of study in both sociology and cultural anthropology are human cultures and cultural variation. However, Dual Inheritance theorists charge that both disciplines too often treat culture as a static superorganic entity that dictates human behavior. Cultures are defined by a suite of common traits shared by a large group of people. DIT theoriests argue that this doesn't sufficiently explain variation in cultural traits at the individual level. By contrast, DIT models human culture at the individual level and views culture as the result of a dynamic evolutionary process at the population level.

Human Sociobiology and Evolutionary Psychology

Human sociobiologists and evolutionary psychologists try to understand how maximizing genetic fitness, in either the modern era or past environments, can explain human behavior. To the human sociobiologist and evolutionary psychologist , culture is either trivial or so bound by genetic fitness that it is unimportant. When faced with a common and seemingly maladaptive trait, practitioners from these disciplines try to determine how the trait actually increases genetic fitness (maybe through kin selection or by speculating about early evolutionary environments). Dual Inheritance theorists, in contrast, will consider a variety of genetic and cultural processes in addition to natural selection on genes.

Human Behavioral Ecology

Human behavioral ecology (HBE) and DIT have a similar relationship to what ecology and evolutionary biology have in the biological sciences. HBE is more concerned about ecological process and DIT more focused on historical process. One difference is that human behavioral ecologists often assume that culture is a system that produces the most adaptive outcome in a given environment. This implies that similar behavioral traditions should be found in similar environments. However, this is not always the case. A study of African cultures showed that cultural history was a better predictor of cultural traits than local ecological conditions.

Memetics

Memetics, which comes from the meme idea described in Dawkins's The Selfish Gene, is similar to DIT in that it treats culture as an evolutionary process that is distinct from genetic transmission. However, there are some philosophical differences between memetics and DIT. One difference is that memetics' focus is on the selection potential of discrete replicators (memes), where DIT allows for transmission of both non-replicators and non-discrete cultural variants. DIT does not assume that replicators are necessary for cumulative adaptive evolution. DIT also more strongly emphasizes the role of genetic inheritance in shaping the capacity for cultural evolution. But perhaps the biggest difference is a difference in academic lineage. Memetics as a label is more influential in popular culture than in academia. Critics of memetics argue that it is lacking in empirical support or is conceptually ill-founded, and question whether there is hope for the memetic research program succeeding. Proponents point out that many cultural traits are discrete, and that many existing models of cultural inheritance assume discrete cultural units, and hence involve memes.

Criticisms

A number of criticisms of DIT have been put forward.From some points of view, use of the term 'dual inheritance' to refer to both what is transmitted genetically and what is transmitted culturally is technically misleading. Some biologists distinguish between inherited information, which is sequestered (encoded in a set of self-assembly instructions and protected from extensive change), and acquired information, which an organism obtains during its lifetime. This, along with debates about terminology, has generated significant resistance to the concept of cultural or behavioral evolution. Many opponents cite horizontal transmission of ideas to be so "different" from the typical vertical transmission (reproduction) in genetic evolution that it is not evolution.

However, 1)even genetic evolution uses non-vertical transmission through the environmental alteration of the genome during life by acquired circumstance: epigenetics, and 2) genetic evolution is also affected by direct horizontal transmission between separate species of plants and strains of bacteria: horizontal gene transfer. Other critics argue that there can be no "dual" inheritance without cultural inheritance being "sequestered" by the biotic genome . Evidence for this process is scarce and controversial. Why this is a demand of critics, however, can be considered unclear as it refutes none of the central claims laid down by proponents of DIT.

More serious criticisms of DIT arise from the choice of Darwinian selection as an explanatory framework for culture. Some argue, cultural evolution does not possess the algorithmic structure of a process that can be modeled in a Darwinian framework as characterized by John von Neumann and used by John Holland to design the genetic algorithm. Forcing culture into a Darwinian framework gives a distorted picture of the process for several reasons. First, some argue Darwinian selection only works as an explanatory framework when variation is randomly generated. To the extent that transmission biases are operative in culture, they mitigate the effect of Darwinian change, i.e. change in the distribution of variants over generations of exposure to selective pressures. Second, since acquired change can accumulate orders of magnitude faster than inherited change, if it is not getting regularly discarded each generation, it quickly overwhelms the population-level mechanism of change identified by Darwin; it 'swamps the phylogenetic signal'. However, DIT proponents might reply that, 1) biotic evolution does not function only on randomly generated phenotypes either, since the phenotypes present in a population are the combined result of random and selective effects during the last generation; and 2)that transmission bias would quite often also reinforce "Darwinian change" since it is widely evidenced that Culture has adaptive value in increasing human fitness.

Another discord in opinion stems from DIT opponents' assertion that there exists some "creative force" that is applied to each idea as it is received and before it is passed on, and that this agency is so powerful that it can be stronger than the selective system of other individuals assessing what to teach and whether your idea has merit . But if this criticism was valid then it would be comparatively much easier to argue an unpopular or incorrect concepts than it actually is. In addition, nothing about DIT runs counter to the idea that an internally selective process (some would call creativity) also determines the fitness of ideas received and sent. In fact this decision making is a large part of the territory embraced by DIT proponents but is poorly understood due to limitations in neurobiology. Related criticisms of the effort to frame culture in Darwinian terms have been leveled by Richard Lewontin , Niles Eldredge, and Stuart Kauffman .

Memetic Engineering

Memetic engineering is a term developed and coined by Leveious Rolando, John Sokol, and Gibran Burchett while they researched and observed the behavior of people after being purposely exposed (knowingly and unknowingly) to certain memetic themes. The term is based on Richard Dawkins' theory of memes.

- The process of developing memes, through meme-splicing and memetic synthesis, with the intent of altering the behavior of others in society or humanity.
- The process of creating and developing theories or ideologies based on an analytical study of societies, cultures, their ways of thinking and the evolution of their minds.
- The process of modifying human beliefs, thought patterns, etc.

Origins of Memetic Engineering

Memetic Engineering developed from diverse influences, including cutting-edge physics of consciousness and memetics research, chaos theory, semiotics, culture jamming, military information warfare, and the viral texts of iconoclasts William S. Burroughs, J.G. Ballard, and Genesis P-Orridge. It draws upon Third Culture sciences and conceptual worldviews for Social Engineering, Values Systems Alignment, and Culture Jamming purposes.

An important example of macro-historical memetic engineering analysis explaining how domination, patriarchy, war and violence are culturally programmed is Riane Eisler's The Chalice and the Blade (San Francisco: Harper SanFrancisco, 1988), which outlines her Dominator and Partnership Culture thesis.

The savvy memetic engineer is able to isolate, study, and subtly manipulate the underlying values systems, symbolic balance and primal atavisms that unconsciously influence the individual psyche and collective identity.

A highly educated but susceptible intelligentsia, worldwide travel, and information vectors like the Internet, cable television, and tabloid media, means that hysterical epidemics and disinformation campaigns may become more common.

This warfare will be conducted using aesthetics, symbols, and doctrines as camouflage that will ultimately influence our cultural meme pool. These contemporary life conditions (Historic Times; Geographic Place; Existential Problems; and Societal Circumstances) are explored in books like Carl Sagan's The Demon Haunted World: Science As a Candle in the Dark (New York: Ballantine Books, 1996), John Brockman's The Third Culture: Beyond the Scientific Revolution (New York: Touchstone Books, 1996), and Michael Shermer's Why People Believe Weird Things: Pseudo-science, Superstition, and Other Confusions of Our Time (New York: W.H. Freeman & Co, 1996).

Fictional descriptions of memetic engineering include Isaac Asimov's seminal Foundation Trilogy (New York: Bantam Books, 1991), George Gurdjieff's artificial mythology Beelzebub's Tales to His Grandson (New York: Penguin USA, 1999); Neil Stephenson's novels Snow Crash (New York: Bantam Spectra, 1993) and The Diamond Age (New York: Bantam Spectra, 1996); and Robert W. Chambers' unearthly The King in Yellow (Buccaneer Books, 1996) tome, which influenced seminal horror author H. P. Lovecraft.

Applied Memetic Engineering

Memetic engineering as a social science lends examples of itself in multiple areas and disciplines. It is currently being examined and researched by the US military as a means to counter insurgency and combat terrorism as explained below in From the Clash to the Confluence of Civilizations by Thomas P.M. Barnett an American military geostrategist, the Pentagon and Richard J. Pech's Inhibiting Imitative Terrorism Through Memetic Engineering.

Other examples of applied memetic engineering are present but not exclusive to the marketing and advertising industries. The question is whether these individuals can be truly considered memetic engineers. Marketing and advertising professionals create memes on an ongoing basis however, this alone doesn't necessarily qualify them as being memetic engineers.

Few if any actually fall into this category. This is possibly due to a lack of understanding of the various memes that have taken root in their target audience minds. According to the definition, industrial designers, music artist, athletes and other entertainers would more likely better serve this definition.

This is because of their ability to create products, phrases and ideas that disseminate the population triggering a response within the brain causing a cultural phenomena. Game theory provides an empirical means of advancing the science of memetics. Memetic game theory, attempts to mathematically capture behavior in strategic situations; where an individual's success in making choices depends on the choices of others, based on past experiences, emotional behavior and learned behavior. It also offers a scientific approach to analyzing social interactions.

According to G.Burchett and his research from G. Edward Griffin's book The Creature from Jekyll Island: A Second Look at the Federal Reserve, the concept of currency or money is considered a meme developed by early bankers.

Here we have an example of a meme that was created to better serve the economic foundation of a society. This is a positive example of memetic engineering. However once the meme of legal tender was created it forced individuals to accept and use the currency under the guise of law. Once an individual is forced to accept currency by the government, essentially the government now controls the means of that individual's survival. In the case of the United States legal tender; the economic control of the citizens was placed in hands of the Federal Reserve which is a private entity.

Psycholinguistics

Psycholinguistics or psychology of language is the study of the psychological and neurobiological factors that enable humans to acquire, use, comprehend and produce language. Initial forays into psycholinguistics were largely philosophical ventures, due mainly to a lack of cohesive data on how the human brain functioned.

Modern research makes use of biology, neuroscience, cognitive science, linguistics, and information theory to study how the brain processes language. There are a number of subdisciplines with non-invasive techniques for studying the neurological workings of the brain; for example, neurolinguistics has become a field in its own right.

Psycholinguistics covers the cognitive processes that make it possible to generate a grammatical and meaningful sentence out of vocabulary and grammatical structures, as well as the processes that make it possible to understand utterances, words, text, etc. Developmental psycholinguistics studies children's ability to learn language.

Areas of Study

Psycholinguistics is an interdisciplinary field. Hence, it is studied by researchers from a variety of different backgrounds, such as psychology, cognitive science, linguistics, and speech and language pathology.

Psycholinguists study many different topics, but these topics can generally be divided into answering the following questions: (1) how do children acquire language (language acquisition)?;

(2) how do people process and comprehend language (language comprehension)?; (3) how do people produce language (language production)?; and (4) how do adults acquire a new language (second language acquisition)? Subdivisions in psycholinguistics are also made based on the different components that make up human language.

Linguistics-related areas:

- Phonetics and phonology are concerned with the study of speech sounds. Within psycholinguistics, research focuses on how the brain processes and understands these sounds.
- Morphology is the study of word structures, especially the relationships between related words (such as dog and dogs) and the formation of words based on rules (such as plural formation).

- Syntax is the study of the patterns which dictate how words are combined to form sentences.
- Semantics deals with the meaning of words and sentences. Where syntax is concerned with the formal structure of sentences, semantics deals with the actual meaning of sentences.
- Pragmatics is concerned with the role of context in the interpretation of meaning.

A researcher interested in language comprehension may study word recognition during reading to examine the processes involved in the extraction of orthographic, morphological, phonological, and semantic information from patterns in printed text. A researcher interested in language production might study how words are prepared to be spoken starting from the conceptual or semantic level. Developmental psycholinguistics study infants' and children's ability to learn and process language.

Theories

Language Acquisition

There are essentially two schools of thought as to how children acquire or learn language, and there is still much debate as to which theory is the correct one. The first theory states that all language must be learned by the child. The second view states that the abstract system of language cannot be learned, but that humans possess an innate language faculty, or an access to what has been called universal grammar. The view that language must be learned was especially popular before 1960 and is well represented by the mentalistic theories of Jean Piaget and the empiricist Rudolf Carnap. Likewise, the school of psychology known as behaviorism by B.F. Skinner) puts forth the point of view that language is a behavior shaped by conditioned response, hence it is learned.

The innatist perspective began with Noam Chomsky's highly critical review of Skinner's book in 1959. This review helped to start what has been termed "the cognitive revolution" in psychology. Chomsky posited humans possess a special, innate ability for language and that complex syntactic features, such as recursion, are "hard-wired" in the brain. These abilities are thought to be beyond the grasp of the most intelligent and social non-humans. According to Chomsky, children acquiring a language have a vast search space to explore among all possible human grammars, yet at the time there was no evidence that children receive sufficient input to learn all the rules of their language. Hence, there must be some other innate mechanism that endows a language ability to humans. Such a language faculty is, according to the innatist theory, what defines human language and makes it different from even the most sophisticated forms of animal communication.

The field of linguistics and psycholinguistics since then has been defined by reactions to Chomsky, pro and con. The pro view still holds that the human ability to use language (specifically the ability to use recursion) is qualitatively different from any sort of animal ability. This ability may have resulted from a favorable mutation or from an adaptation of skills evolved for other purposes. The view that language can be learned has had a recent resurgence inspired by emergentism. This view challenges the "innate" view as scientifically unfalsifiable; that is to say, it can't be tested. With the amount of computer power increasing since the 1980's, researchers have been able to simulate language acquisition using neural network models. These models provide evidence that there may, in fact, be sufficient information contained in the input to learn language, even syntax. If this is true, then an innate mechanism is no longer necessary to explain language acquisition.

Language Comprehension

One question in the realm of language comprehension is how people understand sentences as they read (also known as sentence processing). Experimental research has spawned a number of theories about the architecture and mechanisms of sentence comprehension. Typically these theories are concerned with what types of information contained in the sentence the reader can use to build meaning, and at what point in reading does that information become available to the reader. Issues such as "modular" versus "interactive" processing have been theoretical divides in the field.

A modular view of sentence processing assumes that the stages involved in reading a sentence function independently in separate modules. These modulates have limited interaction with one another. For example, one influential theory of sentence processing, the garden-path theory, states that syntactic analysis takes place first. Under this theory as the reader is reading a sentence, he or she creates the simplest structure possible in order to minimize effort and cognitive load. This is done without any input from semantic analysis or context-dependent information. Hence, in the sentence "The evidence examined by the lawyer turned out to be unreliable," by the time the reader gets to the word "examined" he or she has committed to a reading of the sentence in which the evidence is examining something because it is the simplest parse. This commitment is made despite the fact that it results in an implausible situation; we know that experience that evidence can rarely if ever examine something. Under this "syntax first" theory, semantic information is processed at a later stage. It is only later that the reader will recognize that her or she needs to revise the initial parse into one in which "the evidence" is being examined. In this example, readers typically recognize their misparse by the time they reach "by the lawyer" and must go back and re-parse the sentence. This reanalysis is costly and contributes to slower reading times.

In contrast to a modular account, an interactive theory of sentence processing, such as a constraint-based lexical approach assumes that all available information contained within a sentence can be processed at any time. Under an interactive account, for example, the semantics of a sentence (such as plausibility) can come into play early on in order to help determine the structure of a sentence. Hence, in the sentence above, the reader would be able to make use of plausibility information in order to assume that "the evidence" is being examined instead of doing the examining. There are data to support both modular and interactive accounts; which account is the correct one is still up for debate.

Methodologies

Behavioral Tasks

Many of the experiments conducted in psycholinguistics, especially earlier on, are behavioral in nature. In these types of studies, subjects are presented with linguistic stimuli and asked to perform an action. For example, they may be asked to make a judgment about a word (lexical decision), reproduce the stimulus, or name a visually presented word aloud). Reaction times to respond to the stimuli(usually on the order of milliseconds) and proportion of correct responses are the most often employed measures of performance in behavioral tasks. Such experiments often take advantage of priming effects, whereby a "priming" word or phrase appearing in the experiment can speed up the lexical decision for a related "target" word later.

As an example of how behavioral methods can be used in psycholinguistics research, Fischler (1977) investigated word encoding using the lexical decision task. She asked participants to make decisions about whether two strings of letters were English words. Sometimes the strings would be actual English words requiring a "yes" response), and other times they would be nonwords requiring a "no" response. A subset of the licit words were related semantically (e.g., cat-dog) while others were unrelated (e.g., bread-stem). Fischler found that related word pairs were responded to faster when compared to unrelated word pairs. This facilitation suggests that semantic relatedness can facilitate word encoding.

Eye-movements

Recently, eye tracking has been used to study online language processing. Beginning with Rayner (1978) the importance and informativity of eye-movements during reading was established. Later, Tanenhaus et al. (1995) used the visual-world paradigm to study the cognitive processes related to spoken language. Assuming that eye movements are closely linked to the current focus of attention, language processing can be studied by monitoring eye movements while a subject is presented auditorily with linguistic input.

Neuroimaging

Until the recent advent of non-invasive medical techniques, brain surgery was the preferred way for language researchers to discover how language works in the brain. For example, severing the corpus callosum (the bundle of nerves that connects the two hemispheres of the brain) was at one time a treatment for some forms of epilepsy. Researchers could then study the ways in which the comprehension and production of language were affected by such drastic surgery. Where an illness made brain surgery necessary, language researchers had an opportunity to pursue their research.

Newer, non-invasive techniques now include brain imaging by positron emission tomography (PET); functional magnetic resonance imaging (fMRI); event-related potentials (ERPs) in electroence phalography (EEG) and magnetoencephalography (MEG); and transcranial magnetic stimulation (TMS). Brain imaging techniques vary in their spatial and temporal resolutions (fMRI has a resolution of a few thousand neurons per pixel, and ERP has millisecond accuracy). Each type of methodology presents a set of advantages and disadvantages for studying a particular problem in psycholinguistics.

Computational Modeling

Computational modeling—e.g. the DRC model of reading and word recognition proposed by Coltheart and colleagues—is another methodology. It refers to the practice of setting up cognitive models in the form of executable computer programs. Such programs are useful because they require theorists to be explicit in their hypotheses and because they can be used to generate accurate predictions for theoretical models that are so complex that they render discursive analysis unreliable. Another example of computational modeling is McClelland and Elman's TRACE model of speech perception.

Issues and Areas of Research

Psycholinguistics is concerned with the nature of the computations and processes that the brain undergoes to comprehend and produce language. For example, the cohort model seeks to describe how words are retrieved from the mental lexicon when an individual hears or sees linguistic input.

Recent research using new non-invasive imaging techniques seeks to shed light on just where certain language processes occur in the brain. There are a number of unanswered questions in psycholinguistics, such as whether the human ability to use syntax is based on innate mental structures or emerges from interaction with other humans, and whether some animals can be taught the syntax of human language.

Two other major subfields of psycholinguistics investigate first language acquisition, the process by which infants acquire language, and second language acquisition. In addition, it is much more difficult for adults to acquire second languages than it is for infants to learn their first language (bilingual infants are able to learn both of their native languages easily).

Thus, sensitive periods may exist during which language can be learned readily. A great deal of research in psycholinguistics focuses on how this ability develops and diminishes over time. It also seems to be the case that the more languages one knows, the easier it is to learn more.

The field of aphasiology deals with language deficits that arise because of brain damage. Studies in aphasiology can both offer advances in therapy for individuals suffering from aphasia, and further insight into how the brain processes language.

Spiral Dynamics

Spiral Dynamics is a theory of human development introduced in the 1996 book Spiral Dynamics by Don Beck and Chris Cowan. The book was based on the theory of psychology professor Clare W. Graves. "Spiral Dynamics" is a registered trademark of the National Values Center, Inc.

Overview

Spiral Dynamics argues that human nature is not fixed: humans are able, when forced by life conditions, to adapt to their environment by constructing new, more complex, conceptual models of the world that allow them to handle the new problems. Each new model transcends and includes all previous models. According to Beck and Cowan, these conceptual models are organized around so-called Memes: systems of core values or collective intelligences, applicable to both individuals and entire cultures.

In spiral dynamics, the term Meme refers to a core value system, acting as an organizing principle, which expresses itself through memes (self-propagating ideas, habits, or cultural practices). The prepended and superscripted letter v indicates these are not basic memes but value systems which include them. The colors act as reminders for the Life Conditions and Mind Capacities of each system and alternate between cool and warm colors as a part of the model. Within the model, individuals and cultures do not fall clearly in any single category (color). Each person/culture embodies a mixture of the value patterns, with varying degrees of intensity in each. Spiral Dynamics claims not to be a linear or hierarchical model, although this assertion has been contested.

According to Spiral Dynamics, there are infinite stages of progress and regression over time dependent upon the life circumstances of the person/culture, which are constantly in flux. Attaining higher stages of development is not synonymous with attaining a 'better' or 'more correct' values system. All stages co-exist in both healthy and unhealthy states, whereby any stage of development can lead to undesirable outcomes with respect to the health of the human and social environment.

Memes

First tier Meme levels are focused on different themes for existence, and include almost all of the worldviews, cultures, and mental attitudes up to today. New systems build on adaptations of previous levels and seek to solve problems created by living in those earlier ways. (Attaching concrete examples to these levels of psychological existence is difficult and often misleading because (a) there can be multiple reasons for the same behavior and (b) centralization in a single level regarding all aspects of living is rare. Some advocates of Spiral Dynamics argue that these are ways of thinking about things, not types of people. However, Spiral Dynamics does explicitly define a category of people labeled "Spiral Wizards" who have attained equilibrium around "higher level" Memes.

Beige

- Summary: Archaic-instinctive—survivalistic/automatic/ reflexological
- Time of origin: c. 100,000 BC
- Description: "Express self to meet imperative physiological needs through instincts of Homo sapiens."

Purple

- Summary: Animistic-tribalistic magical-animistic Tribal order
- Time of origin: 50,000 BC
- Description: "Sacrifice to the ways of the elders and customs as one subsumed in group." This is the level of traditional cultures.

Red

- Summary: Egocentric-exploitive power gods/dominionist
- Time of origin: 7000 BC
- Description: "Express self (impulsively) for what self desires without guilt and to avoid shame." Expressed by the mentality of street gangs, Vikings, etc.

Blue

- Summary: Absolutistic-obedience mythic order—purposeful/ authoritarian
- Time of origin: c. 3000 BC
- Description: "Sacrifice self for reward to come through obedience to rightful authority in purposeful Way." Embodied by fundamentalist religions.
- Note: Amber is Ken Wilber's current name for this stage.

Orange

- Summary: Multiplistic-achievist scientific/strategic
- Time of origin: c. 1000 AD on (as early as 600 AD according to Graves and Calhoun)
- Description: "Express self (calculatedly) to reach goals and objectives without rousing the ire of important others." Expressed in the Scientific Revolution and the Industrial Revolution.

Green

- Summary: Relativistic-personalistic—communitarian/ egalitarian

- Time of origin: From 1850 on (surged in early 20th century)
- Description: "Sacrifice self-interest now in order to gain acceptance and group harmony." Expressed in 1960s pluralism and systems theory.

Spiral Dynamics theory sees second tier memes as emerging levels that gradually move away from a focus on subsistence-level concerns of the First Tier, and towards a being-level existence. The existence of two different tiers of psychological development was introduced by the founder Graves. Cowan claims it is possible Graves introduced the tier system as a marketing instrument. Up to today there is no research evidence the two tiers exist. Cowan no longer supports the existence of two tiers but claims the only thing now known about the Yellow and Turquoise systems is that they are more complex versions of Orange and Green. The open-ended theory suggests that the levels Coral and beyond are not yet substantially formed and will solidify as a greater portion of society develops towards those memes.

Yellow

- Summary: Systemic-integrative
- Time of origin: 1950s
- Description: "Express self for what self desires, but to avoid harm to others so that all life, not just own life, will benefit."

Turquoise

- Summary: Holistic
- Time of origin: 1970s
- Description: An integrative system which combines an organism's necessary self-interest with the interests of the communities in which it participates. The theory is still forming.

Development

The colour system was added in the 1970s as a graphic element to decorate training materials used by Cowan and Beck. The term Meme was introduced by Beck and Cowan in Spiral Dynamics where the color language replaced original Graves terminology. Graves had used letter pairs to refer to each level and had not considered any connection with memetics. Beck and Cowan emphasized 'change states' which are part of the Graves theory. They identify landmarks on the transformational path between the levels. Graves' original theory uses a double helix model to show the interrelatedness of an individual's perception of life conditions with their inner neuronal systems, producing a level of psychological existence. This double helix of two interacting forces is referred to as a spiral in 'Spiral Dynamics.

Following the release of their book, Beck and Cowan taught this theory in two courses, SDI and SDII. They ceased their formal working relationship in 1999. Since this break both have continued to offer training courses of various types as well as carrying out consultancy work. Beck spent several years applying the theory of Spiral Dynamics while working with individuals and groups in South Africa involved in the transition from Apartheid for which he received a legislative honor from the state of Texas. Beck has continued to facilitate sessions using the model in both business and politics.

Beck and Ken Wilber (with his integral theory) became interested in each other's work, resulting in Beck developing a branch of spiral dynamics that he calls Spiral Dynamics Integral. This version of the theory uses integral concepts such as the four quadrants. Beck is also associated with the spiritual teacher, Andrew Cohen. This approach continues to be applied and developed in organisational and societal contexts.

Cowan does not subscribe to these developments and promotes a version of the theory which he describes as remaining more faithful to the original research of Clare Graves and extending from it. He continues to use the term 'Spiral Dynamics' to describe his work since he co-created it. With his partner, Natasha Todorovic, he has undertaken work in integrating Spiral Dynamics with NLP and other models, and in developing corporate strategy and practical applications. He has also suggested and additional second tier meme Coral which deals with the development of neurological capacities Each of the external websites listed below promotes the organization of one of the two co-authors.

Pathologies

According to Don Beck and Ken Wilber, each Meme has both healthy and unhealthy versions. The pathologies are sometimes referred to as being "mean" as in "Mean Green Meme" (MGM) or "Mean Orange Meme" (MOM) . As examples, the MOM includes the extremes of capitalism like exploitation, environmental devastation and a general lack of ethics and sensitivity, while the MGM includes performative contradictions like anti-hierarchy, anti-competition, etc. Cowan disputes that any credible evidence exists for the existence of the "Mean Green Meme" and that it is a misrepresentation of the theory. While he recognizes the problems mentioned above, Cowan argues that they exist in other value systems as well, such that attaching them only to the Green Meme is too simplistic. He considers the term "mean" to be inappropriate and a theoretical distortion when questions of adaptation or maladaptation, congruence or ineffectiveness are more to the point. Psychopathology potentially exists at all levels and is a different dimension.

Todorovic offers data and analysis leading to the conclusion that the "Mean Green Meme" conjecture is contradicted by the data. Todorovic further concludes that MGM is a "failure of analysis" based on a variety of misconceptions about spiral dynamics and constitutes an "alarming misdiagnosis" and brands MGM a harmful "form of spiral fundamentalism."

Criticisms of Spiral Dynamics

Critics point out that the model's implications are political as well as developmental and that while the terminology of the theory is self-consciously inclusive, the practical implications of the model can be seen as socially elitist and authoritarian.

In his work on the subject, Beck emphasizes that one of the characteristics of "tier two" individuals, also called "Spiral Wizards", is their ability to make superior decisions for all parties concerned and to manufacture consent for their approaches at lower levels using resonant terms and ideas. In addition to outlining an underlying developmental theory, Spiral Dynamics gives explicit suggestions to these "Wizards" for both consensual and non-consensual management of "lower-tier" individuals. One critic of Spiral Dynamics, Michel Bauwens, has argued that some conceptions of what it means to be "second tier" have come to resemble Nietzsche's idea of the Übermensch. Co author Cowan has publicly dissociated himself from the ideas that are currently being promoted by his ex-partner Beck in conjunction with Ken Wilber.

Predictive Medicine

Predictive medicine is a rapidly emerging field of medicine that entails predicting disease and instituting preventive measures in order to either prevent the disease altogether or significantly decrease its impact upon the patient (such as by preventing mortality or limiting morbidity).

While different prediction methodologies exist, such as genomics, proteomics, and cytomics, the most fundamental way to predict future disease is based on genetics. Although proteomics and cytomics allow for the early detection of disease, much of the time those detect biological markers that exist because a disease process has already started. However, comprehensive genetic testing (such as through the use of DNA arrays or full genome sequencing) allows for the estimation of disease risk years to decades before any disease even exists, or even whether a healthy fetus is at higher risk for developing a disease in adolescence or adulthood. Individuals who are more susceptible to disease in the future can be offered lifestyle advice or medication with the aim of preventing the predicted illness.

Current genetic testing guidelines supported by the health care professionals discourage purely predictive genetic testing of minors until they are competent to understand the relevancy of genetic screening so as to allow them to participate in the decision about whether or not it is appropriate for them. Genetic screening of newborns and children in the field of predictive medicine is deemed appropriate if there is a compelling clinical reason to do so, such as the availability of prevention or treatment as a child that would prevent future disease.

The Goal

The goal of predictive medicine is to predict future disease so that health care professionals and the patient themselves can be proactive in instituting lifestyle modifications and increased physician surveillance, such as bi-annual full body skin exams by a dermatologist or internist if their patient is found to have an increased risk of melanoma or an EKG and cardiology examination by a cardiologist if a patient is found to be at increased risk for a cardiac arrhythmia.

Predictive medicine is intended not for patients but more so for healthy individuals, its purpose being to determine whether susceptibility to a particular disease is increased or not.

A tremendous number of association studies have been published in scientific literature that show associations between specific genetic variants in a person's genetic code and a specific disease.

Association and correlation studies have found that a female individual with a mutation in the BRCA1 gene has a 65% cumulative risk of breast cancer.

Genetic variants in the Factor V gene is associated with an increased tendency to form blood clots, such as deep vein thrombosis (DVTs). Genetics tests are expected to reach the market more quickly than new medicines. Myriad Genetics is already generating revenue from genetic tests for BRCA1 and BRCA2. Predictive medicine changes the fundamental paradigm of medicine from being reactive to being proactive and has the potential to significantly decrease the incidence and prevalence of both common and rare diseases.

Health Benefits

The future of medicine's focus may potentially shift from treating existing diseases, typically late in their progression, to preventing disease before it sets in. Predictive medicine is based on probabilities: while it evaluates susceptibility to diseases, it is not able to predict with 100% certainty that a specific disease will occur. Unlike many preventive interventions that are directed at groups (e.g., immunization programs), predictive medicine is conducted on an individualized basis. For example, glaucoma is a monogenic disease whose early detection can allow to prevent permanent loss of vision. Predictive medicine is expected to be most effective when applied to polygenic multifactorial disease that are prevalent in industrialized countries, such as diabetes mellitus, hypertension, and myocardial infarction. With careful usage, predictive medicine methods such as genetic screens can help diagnose inherited genetic disease caused by problems with a single gene (such as cystic fibrosis) and help early treatment. Some forms of cancer and heart disease are inherited as single-gene diseases and some people in these high-risk families may also benefit from access to genetic tests. As more and more genes associated with increased susceptibility to certain diseases are reported, predictive medicine becomes more useful.

Direct-to-Consumer Genetic Testing

Direct-to-Consumer (DTC) genetic testing enables a consumer to screen his or her own genes without having to go though a health care professional. They can be ordered without the permission of a physician. Variety in DTC tests range from those testing for mutations associated with cystic fibrosis to breast cancer alleles. DTC tests make the applicability of predictive medicine very real and accessible to consumers. Benefits of DTC testing include this accessibility, privacy of genetic information, and promotion of proactive health care. Risks of obtaining DTC testing are the lack of governmental regulation and the interpreting of genetic information without professional counseling.

Mechanism

Transcription

Transcription is the process of creating a complementary RNA copy of a sequence of DNA. Both RNA and DNA are nucleic acids, which use base pairs of nucleotides as a complementary language that can be converted back and forth from DNA to RNA by the action of the correct enzymes. During transcription, a DNA sequence is read by RNA polymerase, which produces a complementary, antiparallel RNA strand. As opposed to DNA replication, transcription results in an RNA complement that includes uracil (U) in all instances where thymine (T) would have occurred in a DNA complement.

Transcription can be explained easily in 4 or 5 steps, each moving like a wave along the DNA.

1. RNA polymerase unwinds/"unzips" the DNA by breaking the hydrogen bonds between complementary nucleotides.
2. RNA nucleotides are paired with complementary DNA bases.
3. RNA sugar-phosphate backbone forms with assistance from RNA polymerase.
4. Hydrogen bonds of the untwisted RNA+DNA helix break, freeing the newly synthesized RNA strand.
5. If the cell has a nucleus, the RNA is further processed (addition of a 3' poly-A tail and a 5' cap) and exits through to the cytoplasm through the nuclear pore complex.

Transcription is the first step leading to gene expression. The stretch of DNA transcribed into an RNA molecule is called a transcription unit and encodes at least one gene. If the gene transcribed encodes a protein, the result of transcription is messenger RNA (mRNA), which will then be used to create that protein via the process of translation. Alternatively, the transcribed gene may encode for either ribosomal RNA (rRNA) or transfer RNA (tRNA), other components of the protein-assembly process, or other ribozymes.

A DNA transcription unit encoding for a protein contains not only the sequence that will eventually be directly translated into the protein (the coding sequence) but also regulatory sequences that direct and regulate the synthesis of that protein. The regulatory sequence before (upstream from) the coding sequence is called the five prime untranslated region (5'UTR), and the sequence following (downstream from) the coding sequence is called the three prime untranslated region (3'UTR).

Transcription has some proofreading mechanisms, but they are fewer and less effective than the controls for copying DNA; therefore, transcription has a lower copying fidelity than DNA replication.

As in DNA replication, DNA is read from $3' \rightarrow 5'$ during transcription. Meanwhile, the complementary RNA is created from the $5' \rightarrow 3'$ direction. This means its 5' end is created first in base pairing. Although DNA is arranged as two antiparallel strands in a double helix, only one of the two DNA strands, called the template strand, is used for transcription. This is because RNA is only single-stranded, as opposed to double-stranded DNA. The other DNA strand is called the coding (lagging) strand, because its sequence is the same as the newly created RNA transcript (except for the substitution of uracil for thymine). The use of only the 3' → 5' strand eliminates the need for the Okazaki fragments seen in DNA replication.

Transcription is divided into 5 stages: pre-initiation, initiation, promoter clearance, elongation and termination.

The gene itself is typically a long stretch of DNA which carries genetic information encoded by genetic code. Every molecule of DNA consists of two strands, each of them having 5' and 3' ends oriented in anti-parallel direction.

The coding strand contains the genetic information while template strand (non-coding strand) serves as a blueprint for the production of RNA. The production of RNA copies of the DNA is called transcription, and is performed by RNA polymerase, which adds one RNA nucleotide at a time to a growing RNA strand.

This RNA is complementary to the template 3' → 5' DNA strand, which is itself complementary to the coding 5' → 3' DNA strand. Therefore, the resulting 5' → 3' RNA strand is identical to the coding DNA strand with the exception that thymines (T) are replaced with uracils (U) in the RNA. A coding DNA strand reading "ATG" is transcribed as "AUG" in RNA.

Transcription in prokaryotes is carried out by a single type of RNA polymerase, which needs DNA sequence called Pribnow box and sigma factor (ó factor) to start transcription.

In eukaryotes, the transcription is done by three types of RNA polymerases, each of them needs special DNA2121 sequence called promoter and a set of DNA-binding proteins - transcription factors to initiate the process. RNA polymerase I is responsible for transcription of rRNA genes, while RNA polymerase II transcribes all protein-coding genes but also some non-coding RNAs (e.g. snRNAs, snoRNAs or long non-coding RNAs) as well. It contains special part called C-terminal domain (CTD) that is rich of serines, which after being phosphorylated accumulate factors necessary for RNA modification and maturation. RNA polymerase III transcribes 5S rRNA and tRNA genes but also some small non-coding RNA genes (e.g. 7SK). Transcription ends on a special sequence called terminator.

Major Steps

Pre-initiation

In eukaryotes, RNA polymerase, and therefore the initiation of transcription, requires the presence of a core promoter sequence in the DNA. Promoters are regions of DNA that promote transcription and, in eukaryotes, are found at -30, -75, and -90 base pairs upstream from the transcription start site (abbreviated to TSS). Core promoters are sequences within the promoter that are essential for transcription initiation. RNA polymerase is able to bind to core promoters in the presence of various specific transcription factors.

The most characterized type of core promoter in eukaryotes is a short DNA sequence known as a TATA box, found 25-30 base pairs upstream from the TSS. The TATA box, as a core promoter, is the binding site for a transcription factor known as TATA-binding protein (TBP), which is itself a subunit of another transcription factor, called Transcription Factor II D (TFIID). After TFIID binds to the TATA box via the TBP, five more transcription factors and RNA polymerase combine around the TATA box in a series of stages to form a preinitiation complex. One transcription factor, DNA helicase, has helicase activity and so is involved in the separating of opposing strands of double-stranded DNA to provide access to a single-stranded DNA template. However, only a low, or basal, rate of transcription is driven by the preinitiation complex alone. Other proteins known as activators and repressors, along with any associated coactivators or corepressors, are responsible for modulating transcription rate.

Thus, preinitiation complex contains: 1. Core Promoter Sequence 2. Transcription Factors 3. DNA Helicase 4. RNA Polymerase 5. Activators and Repressors The transcription preinitiation in archaea is, in essence, homologous to that of eukaryotes, but is much less complex.

The archaeal preinitiation complex assembles at a TATA-box binding site; however, in archaea, this complex is composed of only RNA polymerase II, TBP, and TFB (the archaeal homologue of eukaryotic transcription factor II B (TFIIB)).

Initiation

In bacteria, transcription begins with the binding of RNA polymerase to the promoter in DNA. RNA polymerase is a core enzyme consisting of five subunits: 2 α subunits, 1 β subunit, 1 β' subunit, and 1 ω subunit. At the start of initiation, the core enzyme is associated with a sigma factor that aids in finding the appropriate -35 and -10 base pairs downstream of promoter sequences. When the sigma factor and RNA polymerase combine, they form a holoenzyme. Transcription initiation is more complex in eukaryotes. Eukaryotic RNA polymerase does not directly recognize the core promoter sequences. Instead, a collection of proteins called transcription factors mediate the binding of RNA polymerase and the initiation of transcription. Only after certain transcription factors are attached to the promoter does the RNA polymerase bind to it.

The completed assembly of transcription factors and RNA polymerase bind to the promoter, forming a transcription initiation complex. Transcription in the archaea domain is similar to transcription in eukaryotes.

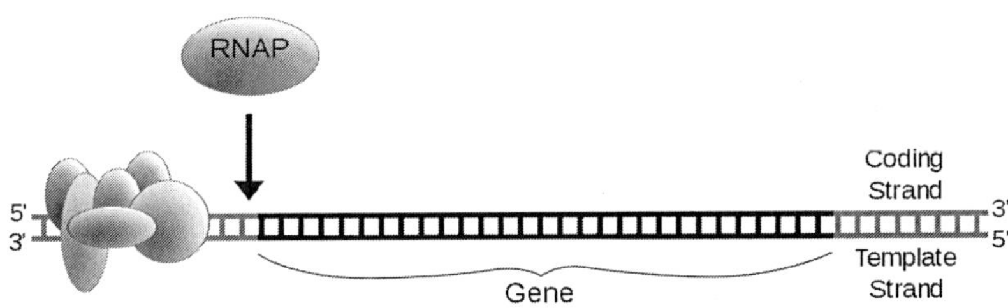

Figure: Simple diagram of transcription initiation. RNAP = RNA polymerase

Promoter Elearance

After the first bond is synthesized, the RNA polymerase must clear the promoter. During this time there is a tendency to release the RNA transcript and produce truncated transcripts. This is called abortive initiation and is common for both eukaryotes and prokaryotes. Abortive initiation continues to occur until the ó factor rearranges, resulting in the transcription elongation complex (which gives a 35 bp moving footprint). The σ factor is released before 80 nucleotides of mRNA are synthesized. Once the transcript reaches approximately 23 nucleotides, it no longer slips and elongation can occur. This, like most of the remainder of transcription, is an energy-dependent process, consuming adenosine triphosphate (ATP). Promoter clearance coincides with phosphorylation of serine 5 on the carboxy terminal domain of RNA Pol in eukaryotes, which is phosphorylated by TFIIH.

Elongation

One strand of the DNA, the template strand (or noncoding strand), is used as a template for RNA synthesis. As transcription proceeds, RNA polymerase traverses the template strand and uses base pairing complementarity with the DNA template to create an RNA copy. Although RNA polymerase traverses the template strand from 3' → 5', the coding (non-template) strand and newly-formed RNA can also be used as reference points, so transcription can be described as occurring 5' → 3'. This produces an RNA molecule from 5' → 3', an exact copy of the coding strand (except that thymines are replaced with uracils, and the nucleotides are composed of a ribose (5-carbon) sugar where DNA has deoxyribose (one less oxygen atom) in its sugar-phosphate backbone).

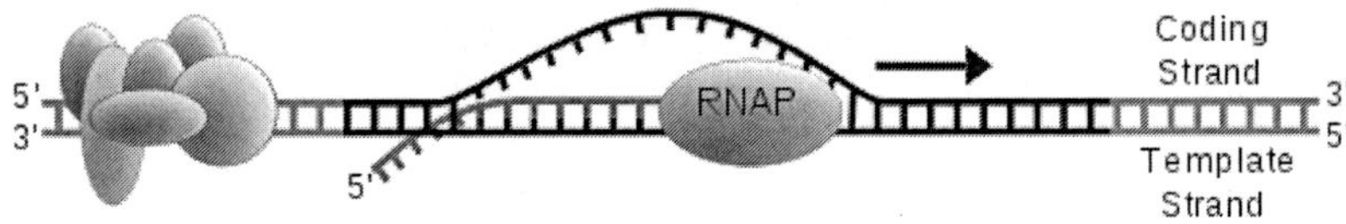

Figure: Simple diagram of transcription elongation

Unlike DNA replication, mRNA transcription can involve multiple RNA polymerases on a single DNA template and multiple rounds of transcription (amplification of particular mRNA), so many mRNA molecules can be rapidly produced from a single copy of a gene. Elongation also involves a proofreading mechanism that can replace incorrectly incorporated bases. In eukaryotes, this may correspond with short pauses during transcription that allow appropriate RNA editing factors to bind. These pauses may be intrinsic to the RNA polymerase or due to chromatin structure.

Termination

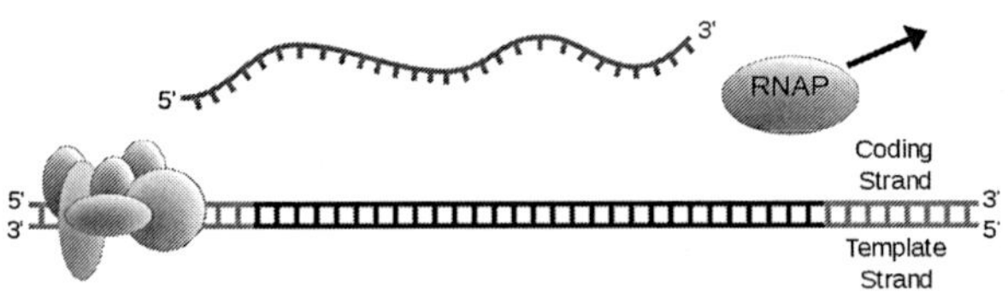

Figure: Simple diagram of transcription termination

Bacteria use two different strategies for transcription termination. In Rho-independent transcription termination, RNA transcription stops when the newly synthesized RNA molecule forms a G-C-rich hairpin loop followed by a run of Us. When the hairpin forms, the mechanical stress breaks the weak rU-dA bonds, now filling the DNA-RNA hybrid. This pulls the poly-U transcript out of the active site of the RNA polymerase, in effect, terminating transcription. In the "Rho-dependent" type of termination, a protein factor called "Rho" destabilizes the interaction between the template and the mRNA, thus releasing the newly synthesized mRNA from the elongation complex.

Transcription termination in eukaryotes is less understood but involves cleavage of the new transcript followed by template-independent addition of As at its new 3' end, in a process called polyadenylation.

Measuring and Detecting Transcription

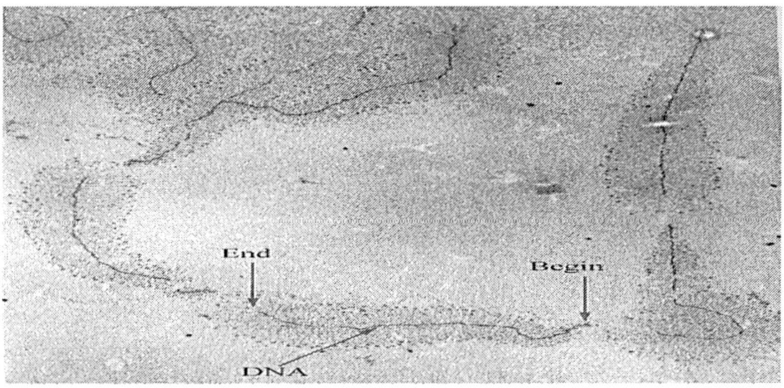

Figure: Electron micrograph of the ribosomal transcription process. The forming mRNA strands are visible as branches from the main DNA strand.

Transcription can be measured and detected in a variety of ways:

- Nuclear Run-on assay: measures the relative abundance of newly formed transcripts
- RNase protection assay and ChIP-Chip of RNAP: detect active transcription sites
- RT-PCR: measures the absolute abundance of total or nuclear RNA levels, which may however differ from transcription rates
- DNA microarrays: measures the relative abundance of the global total or nuclear RNA levels; however, these may differ from transcription rates
- In situ hybridization: detects the presence of a transcript
- MS2 tagging: by incorporating RNA stem loops, such as MS2, into a gene, these become incorporated into newly synthesized RNA. The stem loops can then be detected using a fusion of GFP and the MS2 coat protein, which has a high affinity, sequence-specific interaction with the MS2 stem loops. The recruitment of GFP to the site of transcription is visualised as a single fluorescent spot. This remarkable new approach has revealed that transcription occurs in discontinuous bursts, or pulses . With the notable exception of in situ techniques, most other methods provide cell population averages, and are not capable of detecting this fundamental property of genes.

- Northern blot: the traditional method, and until the advent of RNA-Seq, the most quantitative
- RNA-Seq: applies next-generation sequencing techniques to sequence whole transcriptomes, which allows the measurement of relative abundance of RNA, as well as the detection of additional variations such as fusion genes, post-translational edits and novel splice sites

Transcription Factories

Active transcription units are clustered in the nucleus, in discrete sites called transcription factories or euchromatin. Such sites can be visualized by allowing engaged polymerases to extend their transcripts in tagged precursors (Br-UTP or Br-U) and immuno-labeling the tagged nascent RNA. Transcription factories can also be localized using fluorescence in situ hybridization or marked by antibodies directed against polymerases. There are ~10,000 factories in the nucleoplasm of a HeLa cell, among which are ~8,000 polymerase II factories and ~2,000 polymerase III factories. Each polymerase II factory contains ~8 polymerases. As most active transcription units are associated with only one polymerase, each factory usually contains ~8 different transcription units. These units might be associated through promoters and/or enhancers, with loops forming a 'cloud' around the factor.

History

A molecule that allows the genetic material to be realized as a protein was first hypothesized by François Jacob and Jacques Monod. RNA synthesis by RNA polymerase was established in vitro by several laboratories by 1965; however, the RNA synthesized by these enzymes had properties that suggested the existence of an additional factor needed to terminate transcription correctly.

In 1972, Walter Fiers became the first person to actually prove the existence of the terminating enzyme. Roger D. Kornberg won the 2006 Nobel Prize in Chemistry "for his studies of the molecular basis of eukaryotic transcription".

Reverse Transcription

Some viruses (such as HIV, the cause of AIDS), have the ability to transcribe RNA into DNA. HIV has an RNA genome that is duplicated into DNA. The resulting DNA can be merged with the DNA genome of the host cell. The main enzyme responsible for synthesis of DNA from an RNA template is called reverse transcriptase. In the case of HIV, reverse transcriptase is responsible for synthesizing a complementary DNA strand (cDNA) to the viral RNA genome. An associated enzyme, ribonuclease H, digests the RNA strand, and reverse transcriptase synthesises a complementary strand of DNA to form a double helix DNA structure. This cDNA is integrated into the host cell's genome via another enzyme (integrase) causing the host cell to generate viral proteins that reassemble into new viral particles. In HIV, subsequent to this, the host cell undergoes programmed cell death, apoptosis of T cells. However, in other retroviruses, the host cell remains intact as the virus buds out of the cell.

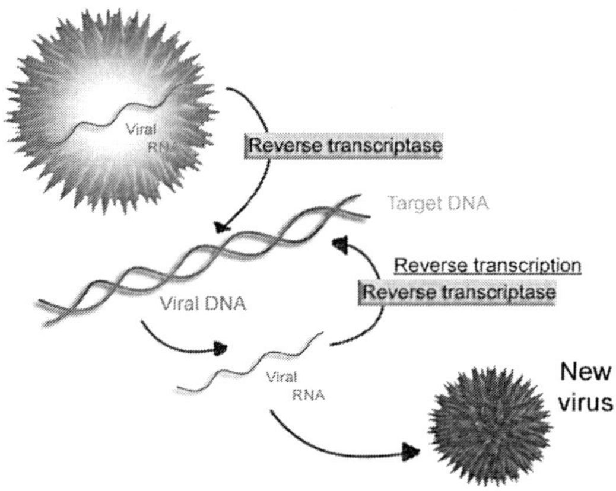

Figure: Scheme of reverse transcription

Some eukaryotic cells contain an enzyme with reverse transcription activity called telomerase. Telomerase is a reverse transcriptase that lengthens the ends of linear chromosomes. Telomerase carries an RNA template from which it synthesizes DNA repeating sequence, or "junk" DNA. This repeated sequence of DNA is important because, every time a linear chromosome is duplicated, it is shortened in length. With "junk" DNA at the ends of chromosomes, the shortening eliminates some of the non-essential, repeated sequence rather than the protein-encoding DNA sequence farther away from the chromosome end. Telomerase is often activated in cancer cells to enable cancer cells to duplicate their genomes indefinitely without losing important protein-coding DNA sequence. Activation of telomerase could be part of the process that allows cancer cells to become immortal. However, the true in vivo significance of telomerase has still not been empirically proven.

RNA Processing

While transcription of prokaryotic protein-coding genes creates messenger RNA (mRNA) which is ready for translation, transcription of eukaryotic genes leaves a primary transcript of RNA (pre-mRNA), which first has to undergo series of modification to become a mature mRNA. These include 5' capping, which is set of enzymatic reactions that add 7-methylguanosine (m G) to the 5' end of pre-mRNA and thus protect the RNA from degradation by exonucleases. The m G cap is then bound by cap binding complex heterodimer (CBC20/CBC80) which aids in mRNA export to cytoplasm and also protect the RNA from decapping. Another modification is 3' cleavage and polyadenylation. They occur if polyadenylation signal sequence (5'- AAUAAA-3') is present in pre-mRNA,which is usually between protein-coding sequence and terminator. The pre-mRNA is first cleaved and then a series of ~200 adenines (A) are added to form poly(A) tail which protects the RNA from degradation. Poly(A) tail is bound by multiple poly(A)-binding proteins (PABP) necessary for mRNA export and translation re-initiation.

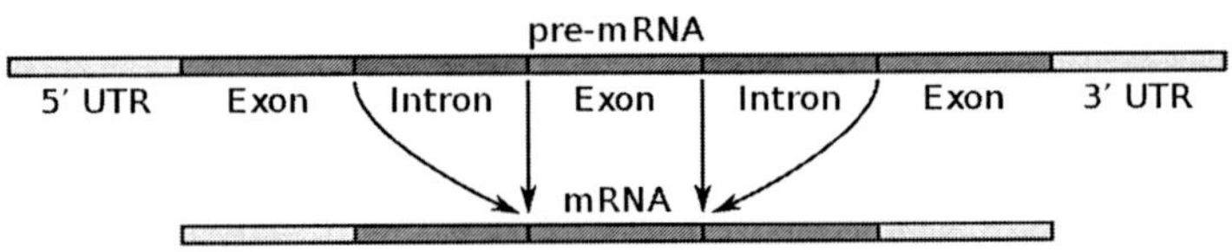

Figure: Simple illustration of exons and introns in pre-mRNA and the formation of mature mRNA by splicing. The UTRs are non-coding parts of exons at the ends of the mRNA.

Very important modification of eukaryotic pre-mRNA is RNA splicing. Majority of eukaryotic pre-mRNAs consist of alternating segments called exons and introns. During the process of splicing, RNA-protein catalytical complex known as spliceosome, catalyze two transesterification reactions, which remove intron and release it in form of lariat structure and then splice neighbouring exons together. In certain cases, some introns or exons can be either removed or retained in mature mRNA.

This so-called alternative splicing creates series of different transcripts originating from a single gene. Because these transcripts can be potentially translated into different proteins, splicing extends the complexity of eukaryotic gene expression. Extensive RNA processing may be an evolutionary advantage made possible by the nucleus of eukaryotes. In prokaryotes transcription and translation happen together whilst in eukaryotes the nuclear membrane separates the two processes giving time for RNA processing to occur.

Post-transcriptional Modification

Post-transcriptional modification is a process in cell biology by which, in eukaryotic cells, primary transcript RNA is converted into mature RNA. A notable example is the conversion of precursor messenger

RNA into mature messenger RNA (mRNA), which includes splicing and occurs prior to protein synthesis. This process is vital for the correct translation of the genomes of eukaryotes as the human primary RNA transcript that is produced as a result of transcription contains both exons, which are coding sections of the primary RNA transcript and introns, which are the non coding sections of the primary RNA transcript.

mRNA Processing

The pre-mRNA molecule undergoes three main modifications. These modifications are 5' capping, 3' polyadenylation, and RNA splicing, which occur in the cell nucleus before the RNA is translated.

5' cap structure

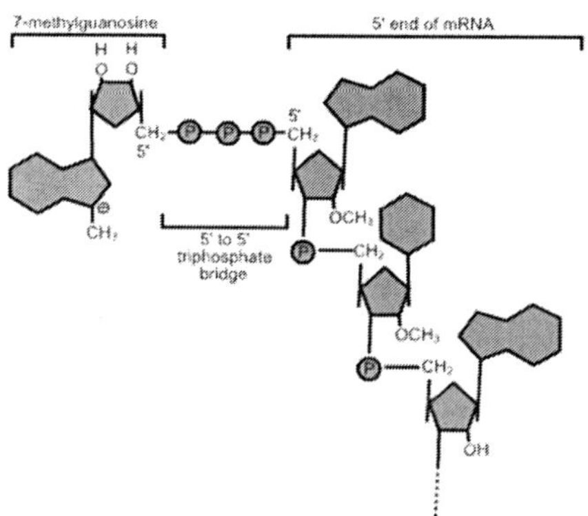

5' Processing

The 5' cap is a specially altered nucleotide on the 5' end of precursor messenger RNA and some other primary RNA transcripts as found in eukaryotes. The process of 5' capping is vital to creating mature messenger RNA, which is then able to undergo translation.

Capping ensures the messenger RNA's stability while it undergoes translation in the process of protein synthesis, and is a highly regulated process that occurs in the cell nucleus.

The 5' cap is found on the 5' end of an mRNA molecule and consists of a guanine nucleotide connected to the mRNA via an unusual 5' to 5' triphosphate linkage.

This guanosine is methylated on the 7 position directly after capping in vivo by a methylated transferase. It is referred to as a 7-methylguanylate cap, abbreviated m G. Further modifications include the possible methylation of the 2' hydroxy-groups of the first 2 ribose sugars of the 5' end of the mRNA. The methylation of both 2' hydroxy-groups is shown on the diagram. The 5' cap looks like the 3' end of an RNA molecule (the 5' carbon of the cap ribose is bonded, and the 3' unbonded). This provides significant resistance to 5' exonucleases.

5' Cap Structure

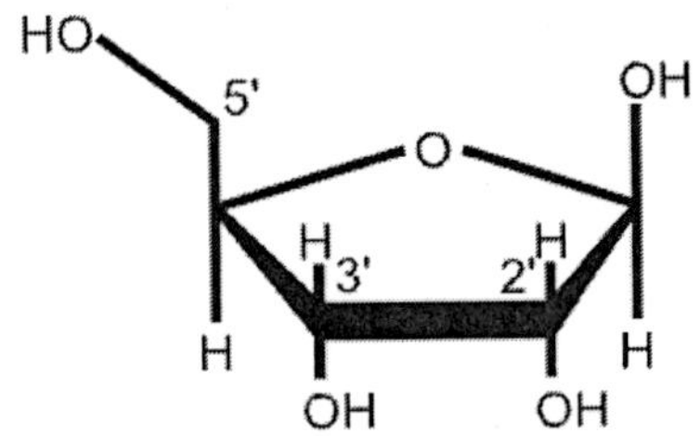

Figure: A Ribose structure showing the positions of the 2', 3' and 5' carbons.

Capping Process

The starting point is the unaltered 5' end of an RNA molecule. This features a final nucleotide followed by three phosphate groups attached to the 5' carbon.

1. One of the terminal phosphate groups is removed (by RNA terminal phosphatase), leaving two terminal phosphates.
2. GTP is added to the terminal phosphates (by a guanylyl transferase), losing two phosphate groups (from the GTP) in the process. This results in the 5' to 5' triphosphate linkage.
3. The 7-nitrogen of guanine is methylated (by a methyl transferase).
4. Other methyltransferases are optionally used to carry out methylation of 5' proximal nucleotides.

5' capping Targeting

The Capping Enzyme Complex (CEC) required for capping is found bound to the RNA polymerase II before transcription starts.

As soon as the 5' end of the new transcript emerges the enzymes transfer to it and begin the capping process (this is a similar kind of mechanism to ensure capping as for polyadenylation). The enzymes for capping can only bind to RNA polymerase II ensuring specificity to only these transcripts, which are almost entirely mRNA.

5' Cap Function

The 5' cap has 4 main functions:

1. Regulation of nuclear export.
2. Prevention of degradation by exonucleases.
3. Promotion of translation .
4. Promotion of 5' proximal intron excision.

Nuclear export of RNA is regulated by the Cap binding complex (CBC), which binds exclusively to capped RNA. The CBC is then recognized by the nuclear pore complex and exported. Once in the cytoplasm after the pioneer round of translation, the CBC is replaced by the translation factors eIF-4E and eIF-4G. This complex is then recognized by other translation initiation machinery including the ribosome.

Cap prevents 5' degradation in two ways. First, degradation of the mRNA by 5' exonucleases is prevented (as mentioned above) by functionally looking like a 3' end. Second, the CBC complex and the eIF-4E/eIF-4G block the access of decapping enzymes to the cap. This increases the half-life of the mRNA, essential in eukaryotes as the export process takes significant time.

Decapping of an mRNA is catalyzed by the decapping complex made up of at least Dcp1 and Dcp2, which must compete with eIF-4E to bind the cap. Thus the 5' cap is a marker of an actively translating mRNA and is used by cells to regulate mRNA half-lives in response to new stimuli. Undesirable mRNAs are sent to P-bodies for temporary storage or decapping, the details of which are still being resolved.

The mechanism of 5' proximal intron excision promotion is not well understood, but the 5' cap appears to loop around and interact with the spliceosome in the splicing process, promoting intron excision.

Capping

Capping of the pre-mRNA involves the addition of 7-methylguanosine (m G) to the 5' end. To achieve this, the terminal 5' phosphate requires removal, which is done with the aid of a phosphatase enzyme. The enzyme guanosyl transferase then catalyses the reaction, which produces the diphosphate 5' end. The diphosphate 5' prime end then attacks the gamma phosphorus atom of a GTP molecule in order to add the guanine residue in a 5'5' triphosphate link. The enzyme (guanine-N -)-methyltransferase ("cap MTase") transfers a methyl group from S-adenosyl methionine to the guanine ring. This type of cap, with just the (m G) in position is called a cap 0 structure. The ribose of the adjacent nucleotide may also be methylated to give a cap 1. Methylation of nucleotides downstream of the RNA molecule produce cap 2, cap 3 structures and so on. In these cases the methyl groups are added to the 2' OH groups of the ribose sugar. The cap protects the 5' end of the primary RNA transcript from attack by ribonucleases that have specificity to the 3'5' phosphodiester bonds. Messenger RNA decapping

The process of messenger RNA decapping consists of hydrolysis of the 5' cap structure on the RNA exposing a 5' monophosphate. This 5' monophosphate is a substrate for the exonuclease Xrn1 and the message is quickly destroyed. There are many situations which may lead to the removal of the cap, some of which are discussed below.

Translation and Decay

Inside cells there is a balance between the processes of translation and mRNA decay. Messages which are being actively translated are bound by polysomes and the initiation factors eIF-4E and eIF-4G. This blocks access to the cap by the decapping enzyme DCP2 and protects the message. In nutrient-starvation conditions or viral infection, translation may be compromised and decapping is stimulated. This balance is reflected in the size and abundance of the cytoplasmic structures known as P-bodies

Specific Decay Pathways

A number of specific decay pathways exist that recognize aberrant messages and promote their decapping. Nonsense mediated decay recognizes premature stop codons and promotes decapping as well as decay by the exosome. Certain classes of miRNA have also been shown to stimulate decapping.

3' Processing

Polyadenylation

Polyadenylation is the addition of a poly(A) tail to an RNA molecule. The poly(A) tail consists of multiple adenosine monophosphates; in other words, it is a stretch of RNA that has only adenine bases. In eukaryotes, polyadenylation is part of the process that produces mature messenger RNA (mRNA) for translation. It, therefore, forms part of the larger process of gene expression.

The process of polyadenylation begins as the transcription of a gene finishes. The 3'-most segment of the newly-made RNA is first cleaved off by a set of proteins; these proteins then synthesize the poly(A) tail at the RNA's 3' end. In some genes, these proteins may add a poly(A) tail at any one of several possible sites. Therefore, polyadenylation can produce more than one transcript from a single gene, similar to alternative splicing.

The poly(A) tail is important for the nuclear export, translation, and stability of mRNA. The tail is shortened over time, and, when it is short enough, the mRNA is enzymatically degraded. However, in a few cell types, mRNAs with short poly(A) tails are stored for later activation by re-polyadenylation in the cytosol. In contrast, when polyadenylation occurs in bacteria, it promotes RNA degradation. This is also sometimes the case for eukaryotic non-coding RNAs. The wide distribution of polyadenylation among living organisms indicates that this process evolved early in the history of life on Earth.

Primer on RNA

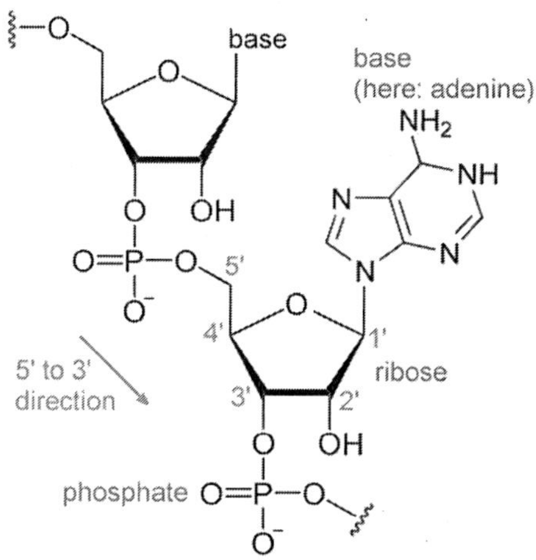

Figure: Chemical structure of RNA. The sequence of bases differs between RNA molecules.

RNAs are a type of large biological molecules, whose individual building blocks are called nucleotides. The name poly(A) tail (for polyadenylic acid tail) reflects the way RNA nucleotides are abbreviated, with a letter for the base the nucleotide contains (A for adenine, C for cytosine, G for guanine and U for uracil). RNAs are produced (transcribed) from a DNA template. By convention, RNA sequences are written in a 5' to 3' direction. The 5' end is the part of the RNA molecule that is transcribed first, and the 3' end is transcribed last. The 3' end is also where the poly(A) tail is found on polyadenylated RNAs.

Messenger RNA (mRNA) is RNA that has a coding region that acts as a template for protein synthesis (translation). The rest of the mRNA, the untranslated regions, tune how active the mRNA is. There are also many RNAs that are not translated, called non-coding RNAs. Like the untranslated regions, many of these non-coding RNAs have regulatory roles.

Nuclear Polyadenylation

Function

In nuclear polyadenylation, a poly(A) tail is added to an RNA at the end of transcription. On mRNAs, the poly(A) tail protects the mRNA molecule from enzymatic degradation in the cytoplasm and aids in transcription termination, export of the mRNA from the nucleus, and translation. Almost all eukaryotic mRNAs are polyadenylated, with the exception of animal replication-dependent histone mRNAs. These are the only mRNAs in eukaryotes that lack a poly(A) tail, ending instead in a stem-loop structure followed by a purine-rich sequence, termed histone downstream element, that directs where the RNA is cut so that the 3' end of the histone mRNA is formed.

Many eukaryotic non-coding RNAs are always polyadenylated at the end of transcription. There are small RNAs where the poly(A) tail is seen only in intermediary forms and not in the mature RNA as the ends are removed during processing, the notable ones being microRNAs. But, for many long noncoding RNAs – a seemingly large group of regulatory RNAs that, for example, includes the RNA Xist, which mediates X chromosome inactivation – a poly(A) tail is part of the mature RNA.

Mechanism

Proteins involved:

CPSF: cleavage/polyadenylation specificity factor

CstF: cleavage stimulation factor

PAP: polyadenylate polymerase

PAB2: polyadenylate binding protein 2

CFI: cleavage factor I

CFII: cleavage factor II

The polyadenylation machinery in the nucleus of eukaryotes works on products of RNA polymerase II, such as precursor mRNA. Here, a multi-protein complex (see components on the right) cleaves the 3'-most part of a newly produced RNA and polyadenylates the end produced by this cleavage. The cleavage is catalysed by the enzyme CPSF and occurs 10–30 nucleotides downstream of its binding site. This site is often the sequence AAUAAA on the RNA, but variants of it that bind more weakly to CPSF exist. Two other proteins add specificity to the binding to an RNA: CstF and CFI. CstF binds to a GU-rich region further downstream of CPSF's site. CFI recognises a third site on the RNA (a set of UGUAA sequences in mammals) and can recruit CPSF even if the AAUAAA sequence is missing. The polyadenylation signal – the sequence motif recognised by the RNA cleavage complex – varies between groups of eukaryotes. Most human polyadenylation sites contain the AAUAAA sequence, but this sequence is less common in plants and fungi.

The RNA is cleaved right after transcription, as CstF also binds to RNA polymerase II. Cleavage also involves the protein CFII, though it is unknown how. The cleavage site associated with a polyadenylation signal can vary up to some 50 nucleotides.

When the RNA is cleaved, polyadenylation starts, catalysed by polyadenylate polymerase. Polyadenylate polymerase builds the poly(A) tail by adding adenosine monophosphate units from adenosine triphosphate to the RNA, cleaving off pyrophosphate. Another protein, PAB2, binds to the new, short poly(A) tail and increases the affinity of polyadenylate polymerase for the RNA. When the poly(A) tail is approximately 250 nucleotides long the enzyme can no longer bind to CPSF and polyadenylation stops, thus determining the length of the poly(A) tail. CPSF is in contact with RNA polymerase II, allowing it to tell the polymerase to terminate transcription. The polyadenylation machinery is also physically linked to the spliceosome, a complex that removes introns from RNAs.

Downstream Effects

The poly(A) tail acts as the binding site for poly(A)-binding protein. Poly(A)-binding protein promotes export from the nucleus and translation, and inhibits degradation. This protein binds to the poly(A) tail prior to mRNA export from the nucleus and in yeast also recruits poly(A) nuclease, an enzyme that shortens the poly(A) tail and allows the export of the mRNA. Poly(A)-binding protein is exported to the cytoplasm with the RNA. mRNAs that are not exported are degraded by the exosome. Poly(A)-binding protein also can bind to, and thus recruit, several proteins that affect translation, one of these is initiation factor-4G, which in turn recruits the 40S ribosomal subunit. However, a poly(A) tail is not required for the translation of all mRNAs.

Deadenylation

In eukaryotic somatic cells, the poly(A) tail of most mRNAs in the cytoplasm gradually get shorter, and mRNAs with shorter poly(A) tail are translated less and degraded sooner. However, it can take many hours before an mRNA is degraded. This deadenylation and degradation process can be accelerated by microRNAs complementary to the 3' untranslated region of an mRNA. In immature egg cells, mRNAs with shortened poly(A) tails are not degraded, but are instead stored without being translated. They are then activated by cytoplasmic polyadenylation after fertilisation, during egg activation.

In animals, poly(A) ribonuclease (PARN) can bind to the 5' cap and remove nucleotides from the poly(A) tail. The level of access to the 5' cap and poly(A) tail is important in controlling how soon the mRNA is degraded. PARN deadenylates less if the RNA is bound by the initiation factors 4E (at the 5' cap) and 4G (at the poly(A) tail), which is why translation reduces deadenylation. The rate of deadenylation may also be regulated by RNA-binding proteins. Once the poly(A) tail is removed, the decapping complex removes the 5' cap, leading to a degradation of the RNA. Several other enzymes that seem to be involved in deadenylation have been identified in yeast.

Alternative Polyadenylation

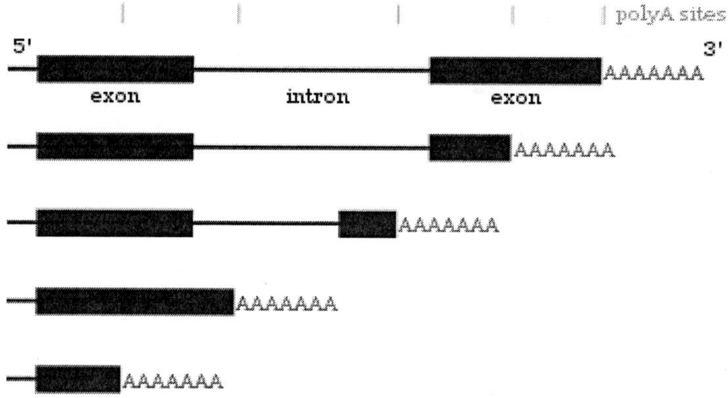

Figure: Results of using different polyadenylation sites on the same gene

Many protein-coding genes have more than one polyadenylation site, so a gene can code for several mRNAs that differ in their 3' end. Since alternative polyadenylation changes the length of the 3' untranslated region, it can change which binding sites for microRNAs the 3' untranslated region contains. MicroRNAs tend to repress translation and promote degradation of the mRNAs they bind to, although there are examples of microRNAs that stabilise transcripts. Alternative polyadenylation can also shorten the coding region, thus making the mRNA code for a different protein, but this is much less common than just shortening the 3' untranslated region.

The choice of poly(A) site can be influenced by extracellular stimuli and depends on the expression of the proteins that take part in polyadenylation. For example, the expression of CstF-64, a subunit of cleavage stimulatory factor (CstF), increases in macrophages in response to lipopolysaccharides (a group of bacterial compounds that trigger an immune response).

This results in the selection of weak poly(A) sites and thus shorter transcripts. This removes regulatory elements in the 3' untranslated regions of mRNAs for defense-related products like lysozyme and TNF-α . These mRNAs then have longer half-lives and produce more of these proteins. RNA-binding proteins other than those in the polyadenylation machinery can also affect whether a polyadenyation site is used,as can DNA methylation near the polyadenylation signal.

Cytoplasmic Polyadenylation

There is polyadenylation in the cytosol of some animal cell types, namely in the germ line, during early embryogenesis and in post-synaptic sites of nerve cells. This lengthens the poly(A) tail of an mRNA with a shortened poly(A) tail, so that the mRNA will be translated. These shortened poly(A) tails are often less than 20 nucleotides, and are lengthened to around 80–150 nucleotides.

In the early mouse embryo, cytoplasmic polyadenylation of maternal RNAs from the egg cell allows the cell to survive and grow even though transcription does not start until the middle of the 2-cell stage (4-cell stage in human).

In the brain, cytoplasmic polyadenylation is active during learning and could play a role in long-term potentiation, which is the strengthening of the signal transmission from a nerve cell to another in response to nerve impulses and is important for learning and memory formation.

Cytoplasmic polyadenylation requires the RNA-binding proteins CPSF and CPEB, and can involve other RNA-binding proteins like Pumilio. Depending on the cell type, the polymerase can be the same type of polyadenylate polymerase (PAP) that is used in the nuclear process, or the cytoplasmic polymerase GLD-2.

Tagging for Degradation in Eukaryotes

For many non-coding RNAs, including tRNA, rRNA, snRNA, and snoRNA, polyadenylation is a way of marking the RNA for degradation, in at least yeast. This polyadenylation is done in the nucleus by the TRAMP complex, which adds a tail that is around 40 nucleotides long to the 3' end. The RNA is then degraded by the exosome. Poly(A) tails have also been found on human rRNA fragments, both the form of homopolymeric (A only) and heterpolymeric (mostly A) tails.

In Prokaryotes and Organelles

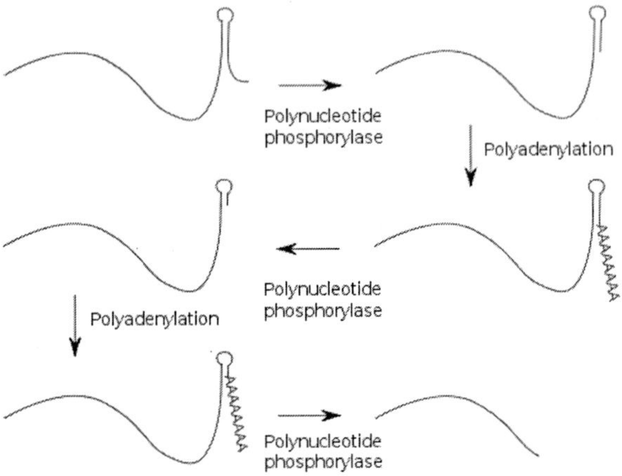

Figure: Polyadenylation in bacteria helps polynucleotide phosphorylase degrade past secondary structure

In many bacteria, both mRNAs and non-coding RNAs can be polyadenylated. This poly(A) tail promotes degradation by the degradosome, which contains two RNA-degrading enzymes: polynucleotide phosphorylase and RNase E. Polynucleotide phosphorylase binds to the 3' end of RNAs and the 3' extension provided by the poly(A) tail allows it to bind to the RNAs whose secondary structure would otherwise block the 3' end. Successive rounds of polyadenylation and degradation of the 3' end by polynucleotide phosphorylase allows the degradosome to overcome these secondary structures. The poly(A) tail can also recruit RNases that cut the RNA in two. These bacterial poly(A) tails are about 30 nucleotides long.

In as different groups as animals and trypanosomes, the mitochondria contain both stabilising and destabilising poly(A) tails. Destabilising polyadenylation targets both mRNA and noncoding RNAs. The poly(A) tails are 43 nucleotides long on average. The stabilising ones start at the stop codon, and without them the stop codon (UAA) is not complete as the genome only encodes the U or UA part. Plant mitochondria have only destabilising polyadenylation, and yeast mitochondria have no polyadenylation at all. While many bacteria and mitochondria have polyadenylate polymerases, they also have another type of polyadenylation, performed by polynucleotide phosphorylase itself. This enzyme is found in bacteria, mitochondria, plastids and as a constituent of the archeal exosome (in those archaea that have an exosome). It can synthesise a 3' extension where the vast majority of the bases are adenines. Like in bacteria, polyadenylation by polynucleotide phosphorylase promotes degradation of the RNA in plastids and likely also archaea.

Evolution

Although polyadenylation is seen in almost all organisms, it is not universal. However, the wide distribution of this modification and the fact that it is present in organisms from all three domains of life implies that the last universal common ancestor of all living organisms, it is presumed, had some form of polyadenylation system. A few organisms do not polyadenylate mRNA, which implies that they have lost their polyadenylation machineries during evolution. Although no examples of eukaryotes that lack polyadenylation are known, mRNAs from the bacterium Mycoplasma gallisepticum and the salt-tolerant archaean Haloferax volcanii lack this modification.

The most ancient polyadenylating enzyme is polynucleotide phosphorylase. This enzyme is part of both the bacterial degradosome and the archaeal exosome, two closely related complexes that recycle RNA into nucleotides. This enzyme degrades RNA by attacking the bond between the 3'-most nucleotides with a phosphate, breaking off a diphosphate nucleotide.

This reaction is reversible, and so the enzyme can also extend RNA with more nucleotides. The heteropolymeric tail added by polynucleotide phosphorylase is very rich in adenine. The choice of adenine is most likely the result of higher ADP concentrations than other nucleotides as a result of using ATP as an energy currency, making it more likely to be incorporated in this tail in early lifeforms. It has been suggested that the involvement of adenine-rich tails in RNA degradation prompted the later evolution of polyadenylate polymerases (the enzymes that produce poly(A) tails with no other nucleotides in them).

Polyadenylate polymerases are not as ancient. They have separately evolved in both bacteria and eukaryotes from CCA-adding enzyme, which is the enzyme that completes the 3' ends of tRNAs. Its catalytic domain is homologous to that of other polymerases. It is presumed that the horizontal transfer of bacterial CCA-adding enzyme to eukaryotes allowed the archaeal-like CCA-adding enzyme to switch function to a poly(A) polymerase. Some lineages, like archaea and cyanobacteria, never evolved a polyadenylate polymerase.

History

Polyadenylation was first identified in 1960 as an enzymatic activity in extracts made from cell nuclei that could polymerise ATP, but not ADP, into polyadenine. Although identified in many types of cells, this activity had no known function until 1971, when poly(A) sequences were found in mRNAs. The only function of these sequences was thought at first to be protection of the 3' end of the RNA from nucleases, but later the specific roles of polyadenylation in nuclear export and translation were identified. The polymerases responsible for polyadenylation were first purified and characterized in the 1960s and 1970s, but the large number of accessory proteins that control this process were discovered only in the early 1990s.

Cleavage and Polyadenylation

The pre-mRNA processing at the 3' end of the RNA molecule involves cleavage of its 3' end and then the addition of about 200 adenine residues to form a poly(A) tail. The cleavage and adenylation reactions occur if a polyadenylation signal sequence (5'- AAUAAA-3') is located near the 3' end of the pre-mRNA molecule, which is followed by another sequence, which is usually (5'-CA-3'). The second signal is the site of cleavage. A GU-rich sequence is also usually present further downstream on the pre-mRNA molecule. After the synthesis of the sequence elements, two multisubunit proteins called cleavage and polyadenylation specificity factor (CPSF) and cleavage stimulation factor (CStF) are transferred from RNA Polymerase II to the RNA molecule. The two factors bind to the sequence elements. A protein complex forms that contains additional cleavage factors and the enzyme Polyadenylate Polymerase (PAP). This complex cleaves the RNA between the polyadenylation sequence and the GU-rich sequence at the cleavage site marked by the (5'-CA-3') sequences. Poly(A) polymerase then adds about 200 adenine units to the new 3' end of the RNA molecule using ATP as a precursor. As the poly(A) tails is synthesised, it binds multiple copies of poly(A) binding protein, which protects the 3'end from ribonuclease digestion.

Splicing

In molecular biology and genetics, splicing is a modification of an RNA after transcription, in which introns are removed and exons are joined. This is needed for the typical eukaryotic messenger RNA before it can be used to produce a correct protein through translation. For many eukaryotic introns, splicing is done in a series of reactions which are catalyzed by the spliceosome, a complex of small nuclear ribonucleoproteins (snRNPs), but there are also self-splicing introns.

RNA splicing is the process by which introns, regions of RNA that do not code for protein, are removed from the pre-mRNA and the remaining exons connected to re-form a single continuous molecule. Although most RNA splicing occurs after the complete synthesis and end-capping of the pre-mRNA, transcripts with many exons can be spliced co-transcriptionally.

The splicing reaction is catalyzed by a large protein complex called the spliceosome assembled from proteins and small nuclear RNA molecules that recognize splice sites in the pre-mRNA sequence. Many pre-mRNAs, including those encoding antibodies, can be spliced in multiple ways to produce different mature mRNAs that encode different protein sequences. This process is known as alternative splicing, and allows production of a large variety of proteins from a limited amount of DNA.

Splicing Pathways

Several methods of RNA splicing occur in nature; the type of splicing depends on the structure of the spliced intron and the catalysts required for splicing to occur.

Spliceosomal Introns

Spliceosomal introns often reside in between eukaryotic protein-coding genes. Within the intron, a 3' splice site, 5' splice site, and branch site are required for splicing. The 5' splice site or splice donor site includes an almost invariant sequence GU at the 5' end of the intron, within a larger, less highly conserved consensus region.

The 3' splice site or splice acceptor site terminates the intron with an almost invariant AG sequence. Upstream (5'-ward) from the AG there is a region high in pyrimidines (C and U), or polypyrimidine tract. Upstream from the polypyrimidine tract is the branch point, which includes an adenine nucleotide. Point mutations in the underlying DNA or errors during transcription can activate a "cryptic splice site" in part of the transcript that usually is not spliced. This results in a mature messenger RNA with a missing section of an exon. In this way a point mutation, which usually only affects a single amino acid, can manifest as a deletion in the final protein.

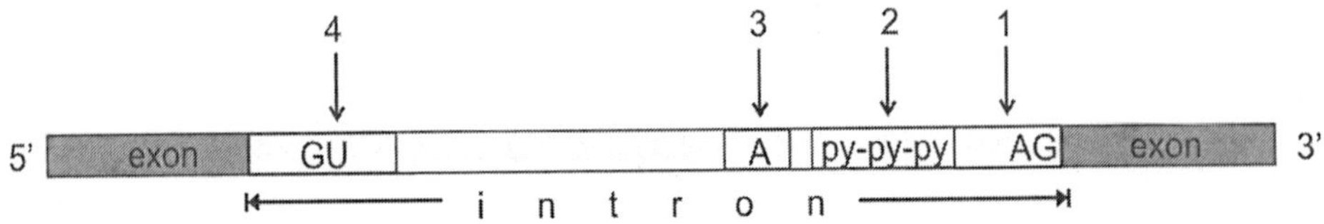

Spliceosome Formation and Activity

Splicing is catalyzed by the spliceosome which is a large RNA-protein complex composed of five small nuclear ribonucleoproteins (snRNPs, pronounced 'snurps').

The RNA components of snRNPs interact with the intron and may be involved in catalysis. Two types of spliceosomes have been identified (the major and minor) which contain different snRNPs.

Major

The major spliceosome splices introns containing GU at the 5' splice site and AG at the 3' splice site.

It is composed of the U1, U2, U4, U5, and U6 snRNPs and is active in the nucleus. In addition, a number of proteins including U2AF and SF1 are required for the assembly of the spliceosome.

- E Complex-U1 binds to the GU sequence at the 5' splice site, along with accessory proteins/enzymes ASF/SF2, U2AF (binds at the Py-AG site), SF1/BBP (BBP=Branch Binding Protein);
- A Complex-U2 binds to the branch site and ATP is hydrolyzed;
- B1 Complex-U5/U4/U6 trimer binds, and the U5 binds exons at the 5' site, with U6 binding to U2;
- B2 Complex-U1 is released, U5 shifts from exon to intron and the U6 binds at the 5' splice site;
- C1 Complex-U4 is released, U6/U2 catalyzes transesterification, that make 5'end of introns ligate to the A on intron and form a lariat ,U5 binds exon at 3' splice site, and the 5' site is cleaved, resulting in the formation of the lariat;
- C2 Complex-U2/U5/U6 remain bound to the lariat, and the 3' site is cleaved and exons are ligated using ATP hydrolysis. The spliced RNA is released and the lariat debranches.

This type of splicing is termed canonical splicing or termed the lariat pathway, which accounts for more than 99% of splicing. By contrast, when the intronic flanking sequences do not follow the GU-AG rule, noncanonical splicing is said to occur (see "minor spliceosome" below).

Minor

The minor spliceosome is very similar to the major spliceosome, however it splices out rare introns with different splice site sequences. While the minor and major spliceosomes contain the same U5 snRNP, the minor spliceosome has different, but functionally analogous snRNPs for U1, U2, U4, and U6, which are respectively called U11, U12, U4atac, and U6atac. Like the major spliceosome, it is only found in the nucleus.

Trans-splicing

Trans-splicing is a form of splicing that joins two exons that are not within the same RNA transcript.

Self-splicing

Self-splicing occurs for rare introns that form a ribozyme, performing the functions of the spliceosome by RNA alone. There are three kinds of self-splicing introns, Group I, Group II and Group III. Group I and II introns perform splicing similar to the spliceosome without requiring any protein. This similarity suggests that Group I and II introns may be evolutionarily related to the spliceosome. Self-splicing may also be very ancient, and may have existed in an RNA world present before protein. Although the two splicing mechanisms described below do not require any proteins to occur, 5 additional RNA molecules and over 50 proteins are used and hydrolyzes many ATP molecules. The splicing mechanisms use ATP in order to accurately splice mRNA's. If the cell were to not use any ATP's, the process would be highly inaccurate and many mistakes would occur.

Two transesterifications characterize the mechanism in which group I introns are spliced:

1. 3'OH of a free guanine nucleoside (or one located in the intron) or a nucleotide cofactor (GMP, GDP, GTP) attacks phosphate at the 5' splice site.
2. 3'OH of the 5'exon becomes a nucleophile and the second transesterification results in the joining of the two exons.

The mechanism in which group II introns are spliced (two transesterification reaction like group I introns) is as follows:

1. The 2'OH of a specific adenosine in the intron attacks the 5' splice site, thereby forming the lariat
2. The 3'OH of the 5' exon triggers the second transesterification at the 3' splice site thereby joining the exons together.

tRNA Splicing

tRNA (also tRNA-like) splicing is another rare form of splicing that usually occurs in tRNA. The splicing reaction involves a different biochemistry than the spliceomsomal and self-splicing pathways. Ribonucleases cleave the RNA and ligases join the exons together.

Evolution

Splicing occurs in all the kingdoms or domains of life, however, the extent and types of splicing can be very different between the major divisions. Eukaryotes splice many protein-coding messenger RNAs and some non-coding RNAs. Prokaryotes, on the other hand, splice rarely and mostly non-coding RNAs. Another important difference between these two groups of organisms is that prokaryotes completely lack the spliceosomal pathway. Because spliceosomal introns are not conserved in all species, there is debate concerning when spliceosomal splicing evolved. Two models have been proposed: the intron late and intron early models.

Splicing Diversity

	Eukaryotes	**Prokaryotes**
Spliceosomal	+	-
Self-splicing	+	+
tRNA	+	+

Biochemical mechanism

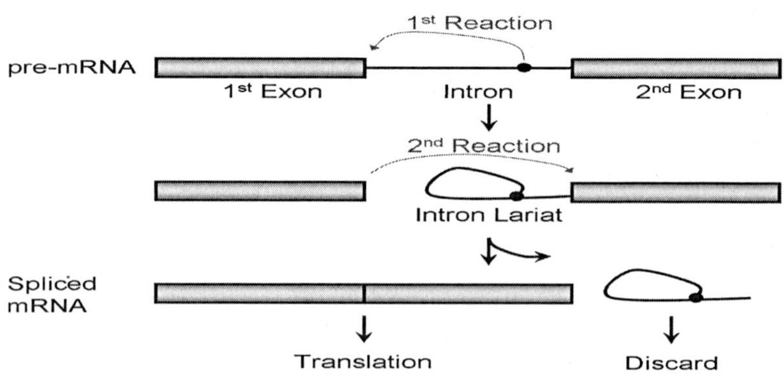

Figure: Diagram illustrating the two-step biochemistry of splicing

Spliceosomal splicing and self-splicing involves a two-step biochemical process. Both steps involve transesterification reactions that occur between RNA nucleotides. tRNA splicing, however, is an exception and does not occur by transesterification.

Spliceosomal and self-splicing transesterification reactions occur via two sequential transesterification reactions. First, the 2'OH of a specific branch-point nucleotide within the intron that is defined during spliceosome assembly performs a nucleophilic attack on the first nucleotide of the intron at the 5' splice site forming the lariat intermediate. Second, the 3'OH of the released 5' exon then performs a nucleophilic attack at the last nucleotide of the intron at the 3' splice site thus joining the exons and releasing the intron lariat.

Alternative Splicing

Alternative splicing (or differential splicing) is a process by which the exons of the RNA produced by transcription of a gene (a primary gene transcript or pre-mRNA) are reconnected in multiple ways during RNA splicing. The resulting different mRNAs may be translated into different protein isoforms; thus, a single gene may code for multiple proteins.

Alternative splicing occurs as a normal phenomenon in eukaryotes, where it greatly increases the diversity of proteins that can be encoded by the genome; in humans, ~95% of multiexonic genes are alternatively spliced. There are numerous modes of alternative splicing observed, of which the most common is exon skipping. In this mode, a particular exon may be included in mRNAs under some conditions or in particular tissues, and omitted from the mRNA in others.

The production of alternatively spliced mRNAs is regulated by a system of trans-acting proteins that bind to cis-acting sites on the pre-mRNA itself. Such proteins include splicing activators that promote the usage of a particular splice site, and splicing repressors that reduce the usage of a particular site. Mechanisms of alternative splicing are highly variable, and new examples are constantly being found, particularly through the use of high-throughput techniques. Researchers hope to fully elucidate the regulatory systems involved in splicing, so that alternative splicing products from a given gene under particular conditions could be predicted by a "splicing code".

Abnormal variations in splicing are also implicated in disease; a large proportion of human genetic disorders result from splicing variants. Abnormal splicing variants are also thought to contribute to the development of cancer, although such aberrant splicing products are, under normal conditions, usually safeguarded and eliminated by a posttranscriptional quality control mechanism. In many cases, the splicing process can create a range of unique proteins by varying the exon composition of the same messenger RNA. This phenomenon is then called alternative splicing. Alternative splicing can occur in many ways. Exons can be extended or skipped, or introns can be retained.

Discovery

Alternative splicing was first observed in 1977. Adenoviruses produce two different primary transcripts, one early in the life cycle and one later, after DNA replication. Researchers found that the primary RNA transcript produced by adenovirus type 2 in the late phase was spliced in different ways, resulting in mRNAs encoding different viral proteins. Both 5' and 3' splice sites varied, and in addition, the transcript contained multiple polyadenylation sites, giving different 3' ends for the processed mRNAs.

In 1981, the first example of alternative splicing in a transcript from a normal, endogenous gene was characterized. The gene encoding the thyroid hormone calcitonin was found to be alternatively spliced in mammalian cells. The pre-mRNA from this gene contains 6 exons; the calcitonin mRNA contains exons 1–4, and terminates after a polyadenylation site in exon 4. Another mRNA is produced from this pre-mRNA by skipping exon 4, and includes exons 1–3, 5, and 6. It encodes a protein known as CGRP (calcitonin gene related peptide). Examples of alternative splicing in immunoglobin gene transcripts in mammals were also observed in the early 1980s.

Since then, alternative splicing has been found to be ubiquitous in eukaryotes. The "record-holder" for alternative splicing is a D. melanogaster gene called Dscam, which could potentially have 38,016 splice variants.

Modes

Five basic modes of alternative splicing are generally recognized.

- Exon skipping or cassette exon: in this case, an exon may be spliced out of the primary transcript or retained. This is the most common mode in mammalian pre-mRNAs.
- Mutually exclusive exons: One of two exons is retained in mRNAs after splicing, but not both.
- Alternative donor site: An alternative 5' splice junction (donor site) is used, changing the 3' boundary of the upstream exon.
- Alternative acceptor site: An alternative 3' splice junction (acceptor site) is used, changing the 5' boundary of the downstream exon.

- Intron retention: A sequence may be spliced out as an intron or simply retained. This is distinguished from exon skipping because the retained sequence is not flanked by introns. If the retained intron is in the coding region, the intron must encode amino acids in frame with the neighboring exons, or a stop codon or a shift in the reading frame will cause the protein to be non-functional. This is the rarest mode in mammals.

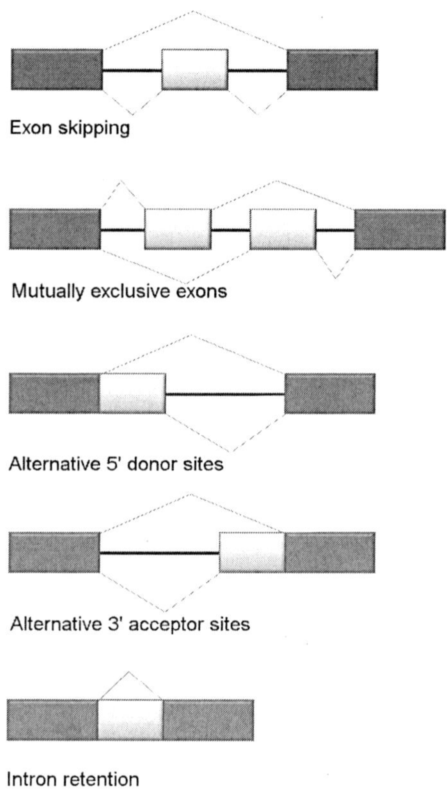

Figure: Traditional classification of basic types of alternative RNA splicing events.

In addition to these primary modes of alternative splicing, there are two other main mechanisms by which different mRNAs may be generated from the same gene; multiple promoters and multiple polyadenylation sites.

Use of multiple promoters is properly described as a transcriptional regulation mechanism rather than alternative splicing; by starting transcription at different points, transcripts with different 5'-most exons can be generated. At the other end, multiple polyadenylation sites provide different 3' end points for the transcript. Both of these mechanisms are found in combination with alternative splicing and provide additional variety in mRNAs derived from a gene.

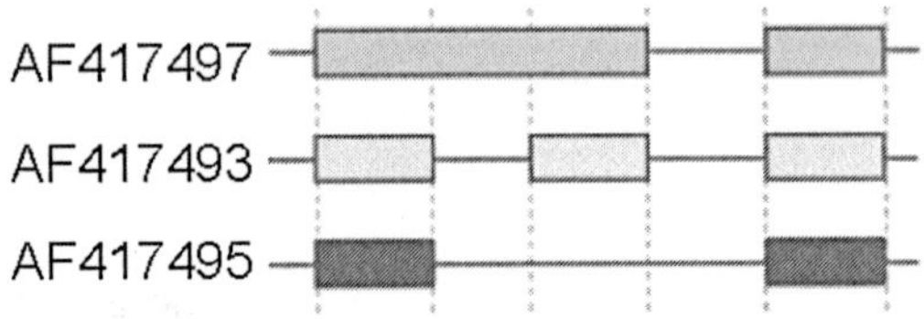

Figure:

Schematic cutoff from 3 splicing structures in the murine hyaluronidase gene. Directionality of transcription from 5' to 3' is shown from left to right. Exons and introns are not drawn to scale.

These modes describe basic splicing mechanisms, but may be inadequate to describe complex splicing events. For instance, the figure to the right shows 3 spliceforms from the mouse hyaluronidase 3 gene. Comparing the exonic structure shown in the first line (green) with the one in the second line (yellow) shows intron retention, whereas the comparison between the second and the third spliceform (yellow vs. blue) exhibits exon skipping. A model nomenclature to uniquely designate all possible splicing patterns has recently been proposed.

Alternative Splicing Mechanisms

General Splicing Mechanism

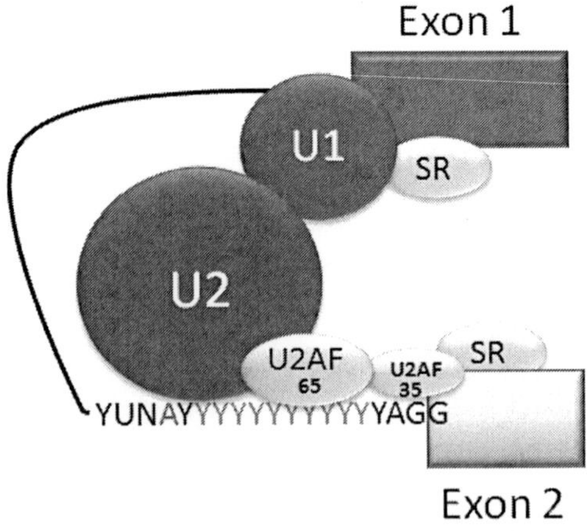

Figure: Spliceosome A complex defines the 5' and 3' ends of the intron before removal

When the pre-mRNA has been transcribed from the DNA, it includes several introns and exons. (In nematodes, the mean is 4–5 exons and introns; in the fruit fly Drosophila there can be more than 100 introns and exons in one transcribed pre-mRNA.) The exons to be retained in the mRNA are determined during the splicing process. The regulation and selection of splice sites are done by trans-acting splicing activator and splicing repressor proteins. The typical eukaryotic nuclear intron has consensus sequences defining important regions. Each intron has GU at its 5' end. Near the 3' end there is a branch site. The nucleotide at the branch point is always an A; the consensus around this sequence varies somewhat. In humans the branch consensus is yUnAy. The branch site is followed by a series of pyrimidines, or polypyrimidine tract, then by AG at 3' end.

Splicing of mRNA is performed by an RNA and protein complex known as the spliceosome, containing snRNPs designated U1, U2, U4, U5, and U6 (U3 is not involved in mRNA splicing). U1 binds to 5' GU and U2 binds to branch site (A) with the assistance of the U2AF protein factors. The complex at this stage is known as the spliceosome A complex. Formation of the A complex is usually the key step in determining the ends of the intron to be spliced out, and defining the ends of the exon to be retained. The U4,U5,U6 complex binds, and U6 replaces the U1 position. U1 and U4 leave. The remaining complex then performs two transeste-rification reactions. In the first transesterification, 5' end of the intron is cleaved from the upstream exon and joined to the branch site A by a 2',5'-phosphodiester linkage. In the second transesterification, the 3' end of the intron is cleaved from the downstream exon, and the two exons are joined by a phosphodiester bond. The intron is then released in lariat form and degraded.

Regulatory Elements and Proteins

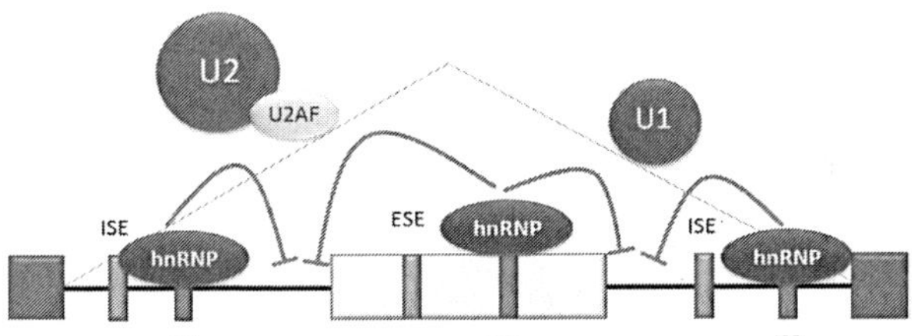

Figure: Splicing repression

Splicing is regulated by trans-acting proteins (repressors and activators), corresponding cis-acting regulatory sites (silencers and enhancers) on the RNA, and other RNA features that influence how splicing will occur, such as secondary structures. Together, these elements form a "splicing code" that governs how splicing will occur under different cellular conditions . The determinants of splicing act in an inter-dependent fashion, but in many cases a particular determinant will have a consistent influence on splicing. Among these cases, there are two major types of cis-acting RNA sequence elements present in pre-mRNAs and they have corresponding trans-acting RNA-binding proteins.

Splicing silencers are sites to which splicing repressor proteins bind, reducing the probability that a nearby site will be used as a splice junction. These can be located in the intron itself (intronic splicing silencers, ISS) or in a neighboring exon (exonic splicing silencers, ESS). They vary in sequence, as well as in the types of proteins that bind to them. The majority of splicing repressors are heterogeneous nuclear ribonucleoproteins (hnRNPs) such as hnRNPA1 and polypyrimidine tract binding protein (PTB).

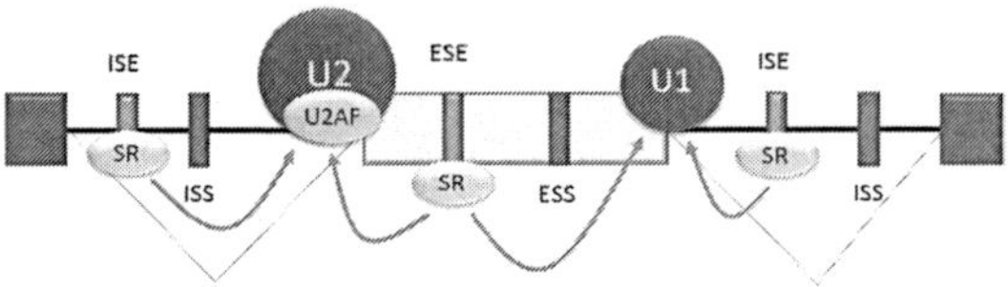

Figure: Splicing Activation

Splicing enhancers are sites to which splicing activator proteins bind, increasing the probability that a nearby site will be used as a splice junction. These also may occur in the intron (intronic splicing enhancers, ISE) or exon (exonic splicing enhancers, ESE). Most of the activator proteins that bind to ISEs and ESEs are members of the SR protein family. Such proteins contain RNA recognition motifs and arginine and serine-rich (RS) domains.

The adaptive significance of splicing silencers and enhancers is attested by a study showing that there is strong selection in human genes against mutations that produce new silencers or disrupt existing enhancers.

In general, the determinants of splicing work in an inter-dependent manner that depends on context, so that the rules governing how splicing is regulated form a splicing code. The presence of a particular cis-acting RNA sequence element may increase the probability that a nearby site will be spliced in some cases, but decrease the probability in other cases, depending on context. The context within which regulatory elements act includes cis-acting context that is established by the presence of other RNA sequence features, and trans-acting context that is established by cellular conditions. For example, some cis-acting RNA sequence elements influence splicing only if multiple elements are present in the same region so as to establish context. As another example, a cis-acting element can have opposite effects on splicing, depending on which proteins are expressed in the cell (e.g., neuronal versus non-neuronal PTB).

Examples

Exon skipping: Drosophila dsx

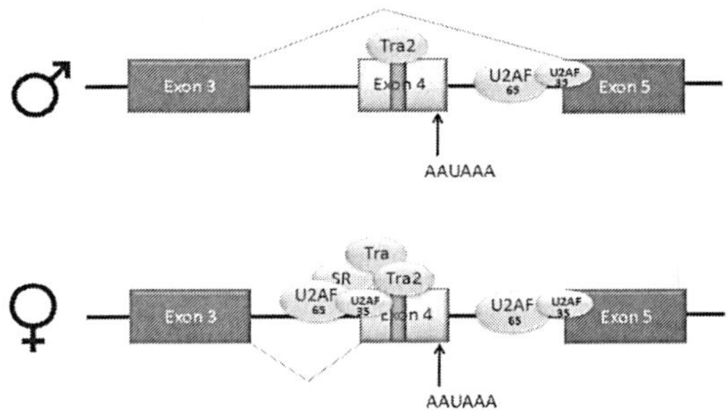

Figure: Alternative splicing of dsx pre-mRNA

Pre-mRNAs from the D. melanogaster gene dsx contain 6 exons. In males, exons 1,2,3,5,and 6 are joined to form the mRNA, which encodes a transcriptional regulatory protein required for male development. In females, exons 1,2,3, and 4 are joined, and a polyadenylation signal in exon 4 causes cleavage of the mRNA at that point. The resulting mRNA is a transcriptional regulatory protein required for female development.

This is an example of exon skipping. The intron upstream from exon 4 has a polypyrimidine tract that doesn't match the consensus sequence well, so that U2AF proteins bind poorly to it without assistance from splicing activators. This 3' splice acceptor site is therefore not used in males. Females, however, produce the splicing activator Transformer (Tra)(see below). The SR protein Tra2 is produced in both sexes and binds to an ESE in exon 4; if Tra is present, it binds to Tra2 and, along with another SR protein, forms a complex that assists U2AF proteins in binding to the weak polypyrimidine tract. U2 is recruited to the associated branch point, and this leads to inclusion of exon 4 in the mRNA.

Alternative acceptor sites: Drosophila Transformer Pre-mRNAs of the Transformer (Tra) gene of Drosophila melanogaster undergo alternative splicing via the alternative acceptor site mode. The gene Tra encodes a protein that is expressed only in females. The primary transcript of this gene contains an intron with two possible acceptor sites.

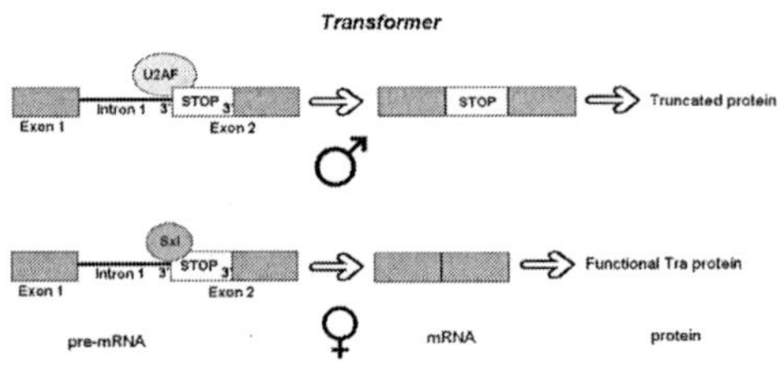

Figure: Alternative splicing of the Drosophila Transformer gene product.

In males, the upstream acceptor site is used. This causes a longer version of exon 2 to be included in the processed transcript, including an early stop codon. The resulting mRNA encodes a truncated protein product that is inactive. Females produce the master sex determination protein Sex lethal (Sxl). The Sxl protein is a splicing repressor that binds to an ISS in the RNA of the Tra transcript near the upstream acceptor site, preventing U2AF protein from binding to the polypyrimidine tract. This prevents the use of this junction, shifting the spliceosome binding to the downstream acceptor site. Splicing at this point bypasses the stop codon, which is excised as part of the intron. The resulting mRNA encodes an active Tra protein, which itself is a regulator of alternative splicing of other sex-related genes (see dsx above).

Exon definition: Fas receptor

Multiple isoforms of the Fas receptor protein are produced by alternative splicing. Two normally occurring isoforms in humans are produced by an exon-skipping mechanism. An mRNA including exon 6 encodes the membrane-bound form of the Fas receptor, which promotes apoptosis, or programmed cell death.

Increased expression of Fas receptor in skin cells chronically exposed to the sun, and absence of expression in skin cancer cells, suggests that this mechanism may be important in elimination of pre-cancerous cells in humans. If exon 6 is skipped, the resulting mRNA encodes a soluble Fas protein that does not promote apoptosis. The inclusion or skipping of the exon depends on two antagonistic proteins, TIA-1 and polypyrimidine tract-binding protein (PTB).

- The 5' donor site in the intron downstream from exon 6 in the pre-mRNA has a weak agreement with the consensus sequence, and is not bound usually by the U1 snRNP. If U1 does not bind, the exon is skipped (see "a" in accompanying figure).
- Binding of TIA-1 protein to an intronic splicing enhancer site stabilizes binding of the U1 snRNP. The resulting 5' donor site

complex assists in binding of the splicing factor U2AF to the 3' splice site upstream of the exon, through a mechanism that is not yet known (see b).

- Exon 6 contains a pyrimidine-rich exonic splicing silencer, ure6, where PTB can bind. If PTB binds, it inhibits the effect of the 5' donor complex on the binding of U2AF to the acceptor site, resulting in exon skipping (see c).

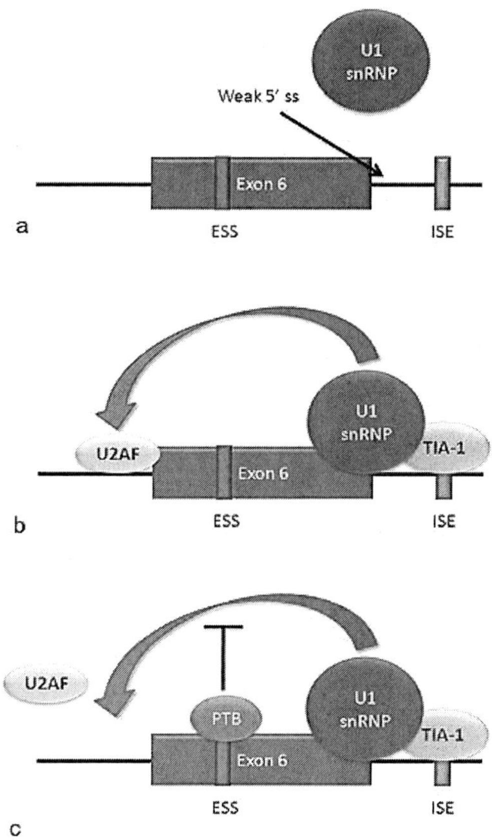

Figure: Alternative splicing of the Fas receptor pre-mRNA

This mechanism is an example of exon definition in splicing. A spliceosome assembles on an intron, and the snRNP subunits bring fold the RNA so that the 5' and 3' ends of the intron are joined.

However, recently studied examples such as this one show that there are also interactions between the ends of the exon. In this particular case, these exon definition interactions are necessary to allow the binding of core splicing factors prior to assembly of the spliceosomes on the two flanking introns. Repressor-activator competition: HIV-1 tat exon 2

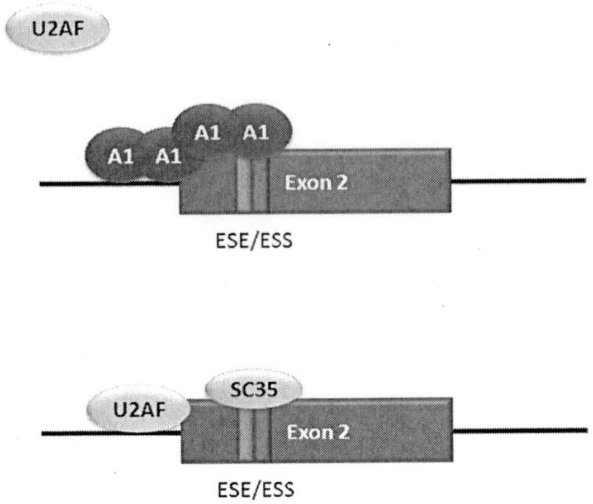

Figure: Alternative splicing of HIV-1 tat exon 2

HIV, the retrovirus that causes AIDS in humans, produces a single primary RNA transcript, which is alternatively spliced in multiple ways to produce over 40 different mRNAs. Equilibrium among differentially spliced transcripts provides multiple mRNAs encoding different products that are required for viral multiplication. One of the differentially spliced transcripts contains the tat gene, in which exon 2 is a cassette exon that may be skipped or included. The inclusion of tat exon 2 in the RNA is regulated by competition between the splicing repressor hnRNP A1 and the SR protein SC35. Within exon 2 an exonic splicing silencer sequence (ESS) and an exonic splicing enhancer sequence (ESE) overlap. If A1 repressor protein binds to the ESS, it initiates cooperative binding of multiple A1 molecules, extending into the 5' donor site upstream of exon 2 and preventing the binding of the core splicing factor U2AF35 to the polypyrimidine tract. If SC35 binds to the ESE, it prevents A1 binding and maintains the 5' donor site in an accessible state for assembly of the spliceosome. Competition between the activator and repressor ensures that both mRNA types (with and without exon 2) are produced.

Adaptive Significance

Alternative splicing is one of several exceptions to the original idea that one DNA sequence codes for one polypeptide (the One gene-one enzyme hypothesis). It might be more correct now to say "One gene – many polypeptides." External information is needed in order to decide which polypeptide is produced, given a DNA sequence and pre-mRNA. Since the methods of regulation are inherited, this provides novel ways for mutations to affect gene expression.

It has been proposed that for eukaryotes alternative splicing was a very important step towards higher efficiency, because information can be stored much more economically. Several proteins can be encoded by a single gene, rather than requiring a separate gene for each, and thus allowing a more varied proteome from a genome of limited size. It also provides evolutionary flexibility.

A single point mutation may cause a given exon to be occasionally excluded or included from a transcript during splicing, allowing production of a new protein isoform without loss of the original protein. Comparative studies indicate that alternative splicing preceded multicellularity in evolution, and suggest that this mechanism might have been co-opted to assist in the development of multicellular organisms.

Research based on the Human Genome Project and other genome sequencing has shown that humans have only about 30% more genes than the roundworm Caenorhabditis elegans, and only about twice as many as the fly Drosophila melanogaster. This finding led to speculation that the perceived greater complexity of humans, or vertebrates generally, might be due to higher rates of alternative splicing in humans than are found in invertebrates.

However, a study on samples of 100,000 ESTs each from human, mouse, rat, cow, fly (D. melanogaster), worm (C. elegans), and the plant Arabidopsis thaliana found no large differences in frequency of alternatively spliced genes among humans and any of the other animals tested. Another study, however, proposed that these results were an artifact of the different numbers of ESTs available for the various organisms. When they compared alternative splicing frequencies in random subsets of genes from each organism, the authors concluded that vertebrates do have higher rates of alternative splicing than invertebrates.

Alternative Splicing and Disease

Changes in the RNA processing machinery may lead to mis-splicing of multiple transcripts, while single-nucleotide alterations in splice sites or cis-acting splicing regulatory sites may lead to differences in splicing of a single gene, and thus in the mRNA produced from a mutant gene's transcripts. A probabilistic analysis indicates that over 60% of human disease-causing mutations affect splicing rather than directly affecting coding sequences.

Abnormally spliced mRNAs are also found in a high proportion of cancerous cells. Until recently, it was unclear whether such aberrant patterns of splicing played a role in causing cancerous growth, or were merely a consequence of cellular abnormalities associated with cancer. It has been shown that there is actually a reduction of alternative splicing in cancerous cells compared to normal ones, and the types of splicing differ; for instance, cancerous cells show higher levels of intron retention than normal cells, but lower levels of exon skipping. Some of the differences in splicing in cancerous cells may result from changes in phosphorylation of trans-acting splicing factors.

Others may be produced by changes in the relative amounts of splicing factors produced; for instance, breast cancer cells have been shown to have increased levels of the splicing factor SF2/ASF. One study found that a relatively small percentage (383 out of over 26000) of alternative splicing variants were significantly higher in frequency in tumor cells than normal cells, suggesting that there is a limited set of genes which, when mis-spliced, contribute to tumor development. It is believed however that the deleterious effects of mis-spliced transcripts are usually safeguarded and eliminated by a cellular posttranscriptional quality control mechanism termed Nonsense-mediated mRNA decay [NMD].

One example of a specific splicing variant associated with cancers is in one of the human DNMT genes. Three DNMT genes encode enzymes that add methyl groups to DNA, a modification that often has regulatory effects. Several abnormally spliced DNMT3B mRNAs are found in tumors and cancer cell lines. In two separate studies, expression of two of these abnormally spliced mRNAs in mammalian cells caused changes in the DNA methylation patterns in those cells. Cells with one of the abnormal mRNAs also grew twice as fast as control cells, indicating a direct contribution to tumor development by this product.

Another example is the Ron (MST1R) proto-oncogene. An important property of cancerous cells is their ability to move and invade normal tissue. Production of an abnormally spliced transcript of Ron has been found to be associated with increased levels of the SF2/ASF in breast cancer cells. The abnomal isoform of the Ron protein encoded by this mRNA leads to cell motility.

Recent provocative studies point to a key function of chromatin structure and histone modifications in alternative splicing regulation. These insights suggest that epigenetic regulation determines not only what parts of the genome are expressed but also how they are spliced.

Genome-wide Analysis of Alternative Splicing

Genome-wide analysis of alternative splicing is a challenging task. Typically, alternatively spliced transcripts have been found by comparing EST sequences, but this requires sequencing of very large numbers of ESTs. Most EST libraries come from a very limited number of tissues, so tissue-specific splice variants are likely to be missed in any case. High-throughput approaches to investigate splicing fall into three categories; DNA microarray-based analyses, CLIP, and in vivo reporter gene assays.

In microarray analysis, arrays of DNA fragments representing individual exons (e.g. Affymetrix exon microarray) or exon/exon boundaries (e.g. arrays from ExonHit or Jivan) have been used. The array is then probed with labeled cDNA from tissues of interest. The probe cDNAs bind to DNA from the exons that are included in mRNAs in their tissue of origin, or to DNA from the boundary where two exons have been joined. This can reveal the presence of particular alternatively spliced mRNAs. Deep sequencing technologies are also being used to perform genome-wide studies of transcript variation.

CLIP (Cross-linking and immunoprecipitation) uses UV radiation to link proteins to RNA molecules in a tissue during splicing. A trans-acting splicing regulatory protein of interest is then precipitated using specific antibodies. When the RNA attached to that protein is isolated and cloned, it reveals the target sequences for that protein.

Finally, it is possible to find the splicing proteins involved in a specific alternative splicing event by constructing reporter genes that will express one of two different fluorescent proteins depending on the splicing reaction that occurs. This method has been used to isolate mutants affecting splicing and thus to identify novel splicing regulatory proteins inactivated in those mutants.

Experimental Manipulation of Splicing

Splicing events can be experimentally altered by binding steric-blocking antisense oligos such as Morpholinos or Peptide nucleic acids to snRNP binding sites, to the branchpoint nucleotide that closes the lariat, or to splice-regulatory element binding sites.

Splicing Errors

Common errors:

- Mutation of a splice site resulting in loss of function of that site. Results in exposure of a premature stop codon, loss of an exon, or inclusion of an intron.

- Mutation of a splice site reducing specificity. May result in variation in the splice location, causing insertion or deletion of amino acids, or most likely, a disruption of the reading frame.
- Displacement of a splice site, leading to inclusion or exclusion of more RNA than expected, resulting in longer or shorter exons.

Many splicing errors are safeguarded by a cellular quality control mechanism termed Nonsense-mediated mRNA decay [NMD].

Protein Splicing

Protein splicing is an intramolecular reaction of a particular protein in which an internal protein segment (called an intein) is removed from a precursor protein with a ligation of C-terminal and N-terminal external proteins (called exteins) on both sides. The splicing junction of the precursor protein is mainly a cysteine or a serine, which are amino acids containing a nucleophilic side chain. The protein splicing reactions which are known now do not require exogenous cofactors or energy sources such as adenosine triphosphate (ATP) or guanosine triphosphate (GTP). Normally, splicing is associated only with pre-mRNA splicing.

In addition to RNA, proteins can undergo splicing. Although the biomolecular mechanisms are different, the principle is the same: parts of the protein, called inteins instead of introns, are removed. The remaining parts, called exteins instead of exons, are fused together. Protein splicing has been observed in a wide range of organisms, including bacteria, archaea, plants, yeast and humans. Types of inteins

The type of the splicing proteins is categorized into four classes: maxi-intein, mini-intein, trans-splicing intein, and alanine intein. The maxi-inteins are N- and C-terminal splicing domains containing an endonuclease domain. The mini-inteins are typical N- and C-terminal splicing domains; however, the endonuclease domain is not present. The trans-splicing inteins are split inteins which are divided into N-termini and C-termini. Alanine inteins have the splicing junction of an alanine instead of a cysteine or a serine, in both of which the protein splicing occurs.

History

Protein splicing was reported by two groups (Anraku and Stevens) in 1990. They both discovered a Saccharomyces cerevisiae VMA1 in a precursor of a vacuolar H-ATPase. The amino acid sequence of the

N- and C-termini corresponded to 70% of that of a vacuolar H -ATPase from other organisms, while the amino acid sequence of the central position corresponded to 30% of that of the yeast HO nuclease

cDNA

In genetics, complementary DNA (cDNA) is DNA synthesized from a messenger RNA (mRNA) template in a reaction catalyzed by the enzyme reverse transcriptase and the enzyme DNA polymerase. cDNA is often used to clone eukaryotic genes in prokaryotes.

When scientists want to express a specific protein in a cell that does not normally express that protein (i.e., heterologous expression), they will transfer the cDNA that codes for the protein to the recipient cell. cDNA is also produced by retroviruses (such as HIV-1, HIV-2, Simian Immunodeficiency Virus, etc.) which is integrated into its host to create a provirus.

Overview

The central dogma of molecular biology outlines that in synthesizing proteins, DNA is transcribed into mRNA, which is translated into protein.

One difference between eukaryotic and prokaryotic genes is that eukaryotic genes can contain introns (intervening sequences) which are not coding sequences (in contrast with exons which are coding sequences), and must be removed from the RNA primary transcript before it becomes mRNA and can be translated into protein. Prokaryotic genes have no introns, so their RNA is not subject to splicing.

Often it is desirable to express eukaryotic genes in prokaryotic cells. A simplified method of doing so would include the addition of eukaryotic DNA to a vector, sometimes a prokaryotic host, which would then transcribe the DNA to mRNA and then translate it to protein.

However, as eukaryotic DNA has introns, and since prokaryotes lack the machinery to splice them, the splicing of eukaryotic DNA must be done prior to adding the eukaryotic DNA into the host. This DNA, which was made as a complementary copy of the RNA and has no introns, is called complementary DNA (cDNA). To obtain expression of the protein encoded by the eukaryotic cDNA, prokaryotic regulatory sequences would also be required (e.g. a promoter).

Synthesis

Though there are several methods for doing so, cDNA is most often synthesized from mature (fully spliced) mRNA using the enzyme reverse transcriptase. This enzyme operates on a single strand of mRNA, generating its complementary DNA based on the pairing of RNA base pairs (A, U, G and C) to their DNA complements (T, A, C and G respectively).

To obtain eukaryotic cDNA whose introns have been removed:

1. A eukaryotic cell transcribes the DNA (from genes) into RNA (pre-mRNA).
2. The same cell processes the pre-mRNA strands by removing introns, and adding a poly-A tail and 5' Methyl-Guanine cap.
3. This mixture of mature mRNA strands is extracted from the cell. The Poly-A tail of the post transcription mRNA can be taken advantage of with oligo(dT) beads in an affinity chromatography assay.
4. A poly-T oligonucleotide primer is hybridized onto the poly-A tail of the mature mRNA template, or random hexamer primers can be added which contain every possible 6 base single strand of DNA and can therefore hybridize anywhere on the RNA (Reverse transcriptase requires this double-stranded segment as a primer to start its operation.)
5. Reverse transcriptase is added, along with deoxynucleoside triphosphates (A, T, G, C). This synthesizes one complementary strand of DNA hybridized to the original mRNA strand.
6. To synthesize an additional DNA strand, you need to digest the RNA of the hybrid strand, using an enzyme like RNase H.
7. After digestion of the RNA, a single stranded DNA (ssDNA) is left and because single stranded nucleic acids are hydrophobic, it tends to loop around itself. It is likely that the ssDNA forms a hairpin loop at the 3' end.
8. From the hairpin loop, a DNA polymerase can then use it as a primer to transcribe a complementary sequence for the ss cDNA.
9. Now, you should be left with a double stranded cDNA with identical sequence as the mRNA of interest.

The reverse transcriptase scans the mature mRNA and synthesizes a sequence of DNA that complements the mRNA template. This strand of DNA is complementary DNA. Note that the central dogma of molecular biology is broken in this process.

Applications

Complementary DNA is often used in gene cloning or as gene probes or in the creation of a cDNA library. When scientists transfer a gene from one cell into another cell in order to express the new genetic material as a protein in the recipient cell, the cDNA will be added to the recipient (rather than the entire gene), because the DNA for an entire gene may include DNA that does not code for the protein or that interrupts the coding sequence of the protein (e.g., introns). Partial sequences of cDNAs are often obtained as expressed sequence tags.

Viruses

Some viruses also use cDNA to turn their viral RNA into mRNA (viral RNA→ cDNA → mRNA). The mRNA is used to make viral proteins to take over the host cell.

Exon

An exon is a nucleic acid sequence that is represented in the mature form of an RNA molecule either after portions of a precursor RNA (introns) have been removed by cis-splicing or when two or more precursor RNA molecules have been ligated by trans-splicing. The mature RNA molecule can be a messenger RNA or a functional form of a non-coding RNA such as rRNA or tRNA. Depending on the context, exon can refer to the sequence in the DNA or its RNA transcript.

History

The term exon derives from expressed region and was coined by American biochemist Walter Gilbert in 1978: "The notion of the cistron... must be replaced by that of a transcription unit containing regions which will be lost from the mature messenger – which I suggest we call introns (for intragenic regions) – alternating with regions which will be expressed – exons."

This definition was originally made for protein-coding transcripts that are spliced before being translated. The term later came to include sequences removed from rRNA and tRNA, and it also was used later for RNA molecules originating from different parts of the genome that are then ligated by trans-splicing.

Function

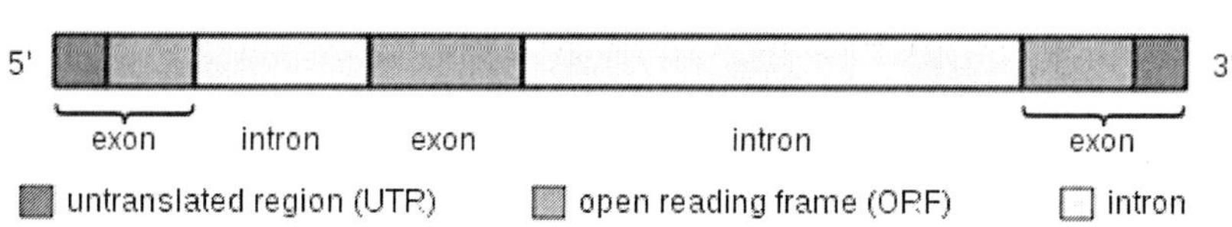

In many genes, each of the exons contains part of the open reading frame (ORF) that codes for a specific portion of the complete protein. However, the term exon is often misused to refer only to coding sequences for the final protein. This is incorrect, since many noncoding exons are known in human genes (Zhang 1998).

To the right is a diagram of a heterogeneous nuclear RNA (hnRNA), which is an unedited mRNA transcript, or pre-mRNAs. Exons can include both sequences that code for amino acids (red) and untranslated sequences (grey). Stretches of unused sequence called introns (blue) are removed, and the exons are joined together to form the final functional mRNA. The notation 5' and 3' refer to the direction of the DNA template in the chromosome and is used to distinguish between the two untranslated regions (grey).

Some of the exons will be wholly or part of the 5' untranslated region (5' UTR) or the 3' untranslated region (3' UTR) of each transcript. The untranslated regions are important for efficient translation of the transcript and for controlling the rate of translation and half-life of the transcript. Furthermore, transcripts made from the same gene may not have the same exon structure, since parts of the mRNA could be removed by the process of alternative splicing. Some mRNA transcripts have exons with no ORFs and, thus, are sometimes referred to as non-coding RNA. Exonization is the creation of a new exon, as result of mutations in intronic sequences. Polycistronic messages have multiple ORFs in one transcript and also have small regions of untranslated sequence between each ORF.

Experimental Approaches that Utilize Exons

Exon trapping or 'gene trapping' is a molecular biology technique that exploits the existence of the intron-exon splicing to find new genes. The first exon of a 'trapped' gene splices into the exon that is contained in the insertional DNA. This new exon contains the ORF for a reporter gene that can now be expressed using the enhancers that control the target gene. A scientist knows that a new gene has been trapped when the reporter gene is expressed.

Splicing can be experimentally modified so that targeted exons are excluded from mature mRNA transcripts by blocking the access of splice-directing small nuclear ribonucleoprotein particles (snRNPs) to pre-mRNA using Morpholino antisense oligos. This has become a standard technique in developmental biology. Morpholino oligos can also be targeted to prevent molecules that regulate splicing (e.g. splice enhancers, splice suppressors) from binding to pre-mRNA, altering patterns of splicing.

Exon Shuffling

Exon shuffling is a theory, introduced by Walter Gilbert in 1977, in which different exons either within a gene or between two nonallelic genes are occasionally mixed. Gilbert suggested that exons might each encode a single protein domain, establishing a kind of modular property. In this fashion, it would be possible for exons to essentially be "mixed and matched" to produce a variety of different proteins, yielded from different combinations of such exons and their resulting domain combinations.

The process of exon shuffling can create a mosaic, or chimeric, protein that is partly built of domains or segments which are similar or identical to domains of other proteins. The mosaic protein is created when an exon from one gene is integrated into another gene's intron. Another kind of exon shuffling is when an exon is duplicated in the same gene.

In the context of evolution, exon shuffling is significant due to its ability to quickly create new multidomain proteins, leading to increased variation in species.

Interrupted Gene: An interrupted gene (also called a split gene) is simply a strand of DNA that contains both introns and exons. Most higher-level eukaryotes have interrupted genes and have longer introns than exons, creating a gene that is longer than its coding region. Interrupted genes are also found in some bacteria. Some eukaryotes, including yeast, have many uninterrupted regions, as they contain long stretches of exons that create necessary mRNA, leading to the development of proteins. This does not mean, however, they are fully uninterrupted, as tRNA synthesis requires excision of a nucleotide sequence, followed by ligation.

Intron

An intron is any nucleotide sequence within a gene that is removed by RNA splicing to generate the final mature RNA product of a gene. The term intron refers to both the DNA sequence within a gene, and the corresponding sequence in RNA transcripts. Sequences that are joined together in the final mature RNA after RNA splicing are exons. Introns are found in the genes of most organisms and many viruses, and can be located in a wide range of genes, including those that generate proteins, ribosomal RNA (rRNA), and transfer RNA (tRNA). When proteins are generated from intron-containing genes, RNA splicing takes place as part of the RNA processing pathway that follows transcription and precedes translation.

The word intron is derived from the term intragenic region, i.e. a region inside a gene. Although introns are sometimes called intervening sequences, the term "intervening sequence" can refer to any of several families of internal nucleic acid sequences that are not present in the final gene product, including inteins, untranslated sequences (UTR), and nucleotides removed by RNA editing, in addition to introns.

Simple Illustration of an Intron and RNA Splicing

Introns and the process of RNA splicing can be illustrated using a text example. Consider the following quotation from Groucho Marx: "Outside of a dog, a book is man's best friend. Inside of a dog it's too dark to read." To make this text look more like a nucleic acid sequence, we'll simply delete all punctuation and make all of the text uppercase. This will be our "message".

1. Identify exon and intron sequences

To remove the intron, the sequences corresponding to the intron and exons must be precisely identified. Here, the exon sequences are changed to bold. Note that misidentifying the boundaries between the exons and intron by even a single letter will result in a nonsensical message.

2. Cleavage at intron-exon junctions

Next, the sequence must be cut precisely at the boundaries between the exons and the intron

3. Link the exons together

Last, the exons must be joined together to generate the final message. (The excised intron is usually broken down to its component letters, which are then recycled.)

Introduction

Introns were first discovered in protein-coding genes of adenovirus, and were subsequently identified in genes encoding transfer RNA and ribosomal RNA genes. Introns are now known to occur within a wide variety of genes throughout organisms and viruses within all of the biological kingdoms. The discovery of introns led to the Nobel Prize in Physiology or Medicine in 1993 for Phillip Allen Sharp and Richard J. Roberts. The term intron was introduced by American biochemist Walter Gilbert:

"The notion of the cistron [...] must be replaced by that of a transcription unit containing regions which will be lost from the mature messenger - which I suggest we call introns (for intragenic regions) - alternating with regions which will be expressed - exons." The frequency of introns within different genomes is observed to vary widely across the spectrum of biological organisms.

For example, introns are extremely common within the nuclear genome of higher vertebrates (e.g. humans and mice), where protein-coding genes almost always contain multiple introns, while introns are rare within the nuclear genes of some eukaryotic microorganisms, for example baker's yeast (Saccharomyces cerevisiae). In contrast, the mitochondrial genomes of vertebrates are entirely devoid of introns, while those of eukaryotic microorganisms may contain many introns. Introns are well known in bacterial and archaeal genes, but occur more rarely than in most eukaryotic genomes.

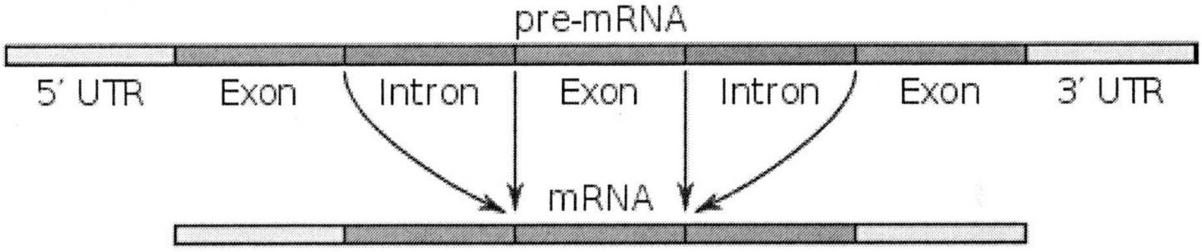

Simple illustration of an unspliced mRNA precursor, with two introns and three exons (top). After the introns have been removed via splicing, the mature mRNA sequence is ready for translation.

Classification

Splicing of all intron-containing RNA molecules is superficially similar, as described above. However, different types of introns were identified through the examination of intron structure by DNA sequence analysis, together with genetic and biochemical analysis of RNA splicing reactions.

At least four distinct classes of introns have been identified.

- Introns in nuclear protein-coding genes that are removed by spliceosomes
- Introns in nuclear and archaeal transfer RNA genes that are removed by proteins (tRNA splicing enzymes)
- Self-splicing group I introns that are removed by RNA catalysis.
- Self-splicing group II introns that are removed by RNA catalysis

Group III introns are proposed to be a fifth family, but little is known about the biochemical apparatus that mediates their splicing. They appear to be related to group II introns, and possibly to spliceosomal introns. Nuclear pre-mRNA introns (spliceosomal introns) are characterized by specific intron sequences located at the boundaries between introns and exons. These sequences are recognized by spliceosomal RNA molecules when the splicing reactions are initiated. In addition, they contain a branch point, a particular nucleotide sequence near the 3' end of the intron that becomes covalently linked to the 5' end of the intron during the splicing process, generating a branched (lariat) intron. Apart from these three short conserved elements, nuclear pre-mRNA intron sequences are highly variable. Nuclear pre-mRNA introns are often much longer than their surrounding exons.

Group I and group II introns are found in genes encoding proteins (messenger RNA), transfer RNA and ribosomal RNA in a very wide range of living organisms., Following transcription into RNA, group I and group II introns also make extensive internal interactions that allow them to fold into a specific, complex three-dimensional architecture. These complex architectures allow some group I and group II introns to be self-splicing, that is, the intron-containing RNA molecule can rearrange its own covalent structure so as to precisely remove the intron and link the exons together in the correct order. In some cases, particular intron-binding proteins are involved in splicing, acting in such a way that the assist the intron in folding into the three-dimensional structure that is necessary for self-splicing activity. Group I and group II introns are distinguished by different sets of internal conserved sequences and folded structures, and by the fact that splicing of RNA molecules containing group II introns generates branched introns (like those of spliceosomal RNAs), while group I introns use a non-encoded guanosine nucleotide (typically GTP) to initiate splicing, adding it on to the 5'-end of the excised intron.

Transfer RNA introns that depend upon proteins for removal occur at a specific location within the anticodon loop of unspliced tRNA precursors, and are removed by a tRNA splicing endonuclease. The exons are then linked together by a second protein, the tRNA splicing ligase. Note that self-splicing introns are also sometimes found within tRNA genes.

2

Overarching Structure

Computational Methods for Studying Protein Folding

Protein Structure Prediction

Protein structure prediction is the prediction of the three-dimensional structure of a protein from its amino acid sequence — that is, the prediction of its secondary, tertiary, and quaternary structure from its primary structure. Structure prediction is fundamentally different from the inverse problem of protein design. Protein structure prediction is one of the most important goals pursued by bioinformatics and theoretical chemistry; it is highly important in medicine (for example, in drug design) and biotechnology (for example, in the design of novel enzymes). Every two years, the performance of current methods is assessed in the CASP experiment (Critical Assessment of Techniques for Protein Structure Prediction).

Secondary Structure

Secondary structure prediction is a set of techniques in bioinformatics that aim to predict the local secondary structures of proteins and RNA sequences based only on knowledge of their primary structure — amino acid or nucleotide sequence, respectively. For proteins, a prediction consists of assigning regions of the amino acid sequence as likely alpha helices, beta strands (often noted as "extended" conformations), or turns. The success of a prediction is determined by comparing it to the results of the DSSP algorithm applied to the crystal structure of the protein; for nucleic acids, it may be determined from the hydrogen bonding pattern. Specialized algorithms have been developed for the detection of specific well-defined patterns such as transmembrane helices and coiled coils in proteins, or canonical microRNA structures in RNA. The best modern methods of secondary structure prediction in proteins reach about 80% accuracy ; this high accuracy allows the use of the predictions in fold recognition and ab initio protein structure prediction, classification of structural motifs, and refinement of sequence alignments. The accuracy of current protein secondary structure prediction methods is assessed in weekly benchmarks such as LiveBench and EVA.

Background

Early methods of secondary structure prediction, introduced in the 1960s and early 1970s, focused on identifying likely alpha helices and were based mainly on helix-coil transition models. Significantly more accurate predictions that included beta sheets were introduced in the 1970s and relied on statistical assessments based on probability parameters derived from known solved structures. These methods, applied to a single sequence, are typically at most about 60-65% accurate, and often underpredict beta sheets. The evolutionary conservation of secondary structures can be exploited by simultaneously assessing many homologous sequences in a multiple sequence alignment, by calculating the net secondary structure propensity of an aligned column of amino acids. In concert with larger databases of known protein structures and modern machine learning methods such as neural nets and support vector machines, these methods can achieve up 80% overall accuracy in globular proteins. The theoretical upper limit of accuracy is around 90%, partly due to idiosyncrasies in DSSP assignment near the ends of secondary structures, where local conformations vary under native conditions but may be forced to assume a single conformation in crystals due to packing constraints. Limitations are also imposed by secondary structure prediction's inability to account for tertiary structure; for example, a sequence predicted as a likely helix may still be able to adopt a beta-strand conformation if it is located within a beta-sheet region of the protein and its side chains pack well with their neighbors. Dramatic conformational changes related to the protein's function or environment can also alter local secondary structure.

Chou-Fasman Method

The Chou–Fasman method are an empirical technique for the prediction of secondary structures in proteins, originally developed in the 1970s. The method is based on analyses of the relative frequencies of each amino acid in alpha helices, beta sheets, and turns based on known protein structures solved with X-ray crystallography. From these frequencies a set of probability parameters were derived for the appearance of each amino acid in each secondary structure type, and these parameters are used to predict the probability that a given sequence of amino acids would form a helix, a beta strand, or a turn in a protein. The method is at most about 50–60% accurate in identifying correct secondary structures, which is significantly less accurate than the modern machine learning–based techniques.

The Chou-Fasman method was among the first secondary structure prediction algorithms developed and relies predominantly on probability parameters determined from relative frequencies of each amino acid's appearance in each type of secondary structure. The original Chou-Fasman parameters, determined from the small sample of structures solved in the mid-1970s, produce poor results compared to modern methods, though the parameterization has been updated since it was first published. The Chou-Fasman method is roughly 50-60% accurate in predicting secondary structures. Amino acid propensities

The original Chou–Fasman parameters found some strong tendencies among individual amino acids to prefer one type of secondary structure over others. Alanine, glutamate, leucine, and methionine were identified as helix formers, while proline and glycine, due to the unique conformational properties of their peptide bonds, commonly end a helix. The original Chou–Fasman parameters were derived from a very small and non-representative sample of protein structures due to the small number of such structures that were known at the time of their original work. These original parameters have since been shown to be unreliable and have been updated from a current dataset, along with modifications to the initial algorithm.

The Chou–Fasman method takes into account only the probability that each individual amino acid will appear in a helix, strand, or turn. Unlike the more complex GOR method, it does not reflect the conditional probabilities of an amino acid to form a particular secondary structure given that its neighbors already possess that structure. This lack of cooperativity increases its computational efficiency but decreases its accuracy, since the propensities of individual amino acids are often not strong enough to render a definitive prediction.

Algorithm

The Chou–Fasman method predicts helices and strands in a similar fashion, first searching linearly through the sequence for a "nucleation" region of high helix or strand probability and then extending the region until a subsequent four-residue window carries a probability of less than 1. As originally described, four out of any six contiguous amino acids were sufficient to nucleate helix, and three out of any contiguous five were sufficient for a sheet. The probability thresholds for helix and strand nucleations are constant but not necessarily equal; originally

1.03 was set as the helix cutoff and 1.00 for the strand cutoff. Turns are also evaluated in four-residue windows, but are calculated using a multi-step procedure because many turn regions contain amino acids that could also appear in helix or sheet regions. Four-residue turns also have their own characteristic amino acids; proline and glycine are both common in turns. A turn is predicted only if the turn probability is greater than the helix or sheet probabilities and a probability value based on the positions of particular amino acids in the turn exceeds a predetermined threshold. The turn probability p(t) is determined as:

$$p(t) = p_t(j) \times p_t(j+1) \times p_t(j+2) \times p_t(j+3)$$

where j is the position of the amino acid in the four-residue window. If p(t) exceeds an arbitrary cutoff value (originally 7.5e–3), the mean of the p(j)'s exceeds 1, and p(t) exceeds the alpha helix and beta sheet probabilities for that window, then a turn is predicted. If the first two conditions are met but the probability of a beta sheet p(b) exceeds p(t), then a sheet is predicted instead.

GOR Method

The GOR method (Garnier-Osguthorpe-Robson) is an information theory-based method for the prediction of secondary structures in proteins. It was developed in the late 1970s shortly after the simpler Chou-Fasman method. Like Chou-Fasman, the GOR method is based on probability parameters derived from empirical studies of known protein tertiary structures solved by X-ray crystallography. However, unlike Chou-Fasman, the GOR method takes into account not only the propensities of individual amino acids to form particular secondary structures, but also the conditional probability of the amino acid to form a secondary structure given that its immediate neighbors have already formed that structure. The method is therefore essentially Bayesian in its analysis.

The GOR method analyzes sequences to predict alpha helix, beta sheet, turn, or random coil secondary structure at each position based on 17-amino-acid sequence windows. The original description of the method included four scoring matrices of size 17×20, where the columns correspond to the log-odds score, which reflects the probability of finding a given amino acid at each position in the 17-residue sequence. The four matrices reflect the probabilities of the central, ninth amino acid being in a helical, sheet, turn, or coil conformation. In subsequent revisions to the method, the turn matrix was eliminated due to the high variability of sequences in turn regions (particularly over such a large window). The method was considered as best requiring at least four contiguous residues to score as alpha helices to classify the region as helical, and at least two contiguous residues for a beta sheet. The mathematics and algorithm of the GOR method were based on an earlier series of studies by Robson and colleagues reported mainly in the Journal of Molecular Biology (e.g.) and The Biochemical Journal (e.g.). The latter describes the information theoretic expansions in terms of conditional information measures. The use of the word "simple" in the title of the GOR paper reflected the fact that the above earlier methods provided proofs and techniques somewhat daunting by being rather unfamiliar in protein science in the early 1970s; even Bayes methods were then unfamiliar and controversial. An important feature of these early studies, which survived in the GOR method, was the treatment of the sparse protein sequence data of the early 1970s by expected information measures.

That is, expectations on a Bayesian basis considering the distribution of plausible information measure values given the actual frequencies (numbers of observations). The expectation measures resulting from integration over this and similar distributions may now be seen as composed of "incomplete" or extended zeta functions, e.g. z(s,observed frequency) " z(s,expected frequency) with incomplete zeta function z (s, n) = 1 + (1/2) + (1/3) + (1/4) + ... + (1/n) . The GOR method used s = 1. Also, in the GOR method and the earlier methods, the measure for the contrary state to e.g. helix H, i.e. H, was subtracted from that for H, and similarly for beta sheet, turns, and coil or loop. Thus the method can be seen as employing a zeta function estimate of log predictive odds. An adjustable decision constant could also be applied, which thus also implies a decision theory approach; the GOR method allowed the option to use decision constants to optimize predictions for different classes of protein.

The expected information measure used as a basis for the information expansion was less important by the time of publication of the GOR method because protein sequence data became more plentiful, at least for the terms considered at that time. Then, for s=1, the expression z(s,observed frequency) " z(s,expected frequency) approaches the natural logarithm of (observed frequency / expected frequency) as frequencies increase. However, this measure (including use of other values of s) remains important in later more general applications with high dimensional data, where data for more complex terms in the information expansion are inevitably sparse (e.g.). The GOR method, named for the three scientists who developed it — Garnier, Osguthorpe, and Robson — is an information theory-based method developed not long after Chou-Fasman. It uses a more powerful probabilistic techniques of Bayesian inference. The method is a specific optimized application of mathematics and algorithms developed in a series of papers by Robson and colleagues, eg. and). The GOR method is capable of continued extension by such principles, and has gone through several versions.

The GOR method takes into account not only the probability of each amino acid having a particular secondary structure, but also the conditional probability of the amino acid assuming each structure given the contributions of its neighbors (it does not assume that the neighbors have that same structure). The approach is both more sensitive and more accurate than that of Chou and Fasman because amino acid structural propensities are only strong for a small number of amino acids such as proline and glycine. Weak contributions from each of many neighbors can add up to strong effect overall. The original GOR method was roughly 65% accurate and is dramatically more successful in predicting alpha helices than beta sheets, which it frequently mispredicted as loops or disorganized regions. Later GOR methods considered also pairs of amino acids, significantly improving performance. The major difference from the following technique is perhaps that the weights in an implied network of contributing terms are assigned a priori, from statistical analysis of proteins of known structure, not by feedback to optimize agreement with a training set of such.

Machine Learning

Neural network methods use training sets of solved structures to identify common sequence motifs associated with particular arrangements of secondary structures. These methods are over 70% accurate in their predictions, although beta strands are still often underpredicted due to the lack of three-dimensional structural information that would allow assessment of hydrogen bonding patterns that can promote formation of the extended conformation required for the presence of a complete beta sheet.

Support vector machines have proven particularly useful for predicting the locations of turns, which are difficult to identify with statistical methods. The requirement of relatively small training sets has also been cited as an advantage to avoid overfitting to existing structural data. Extensions of machine learning techniques attempt to predict more fine-grained local properties of proteins, such as backbone dihedral angles in unassigned regions. Both SVMs and neural networks have been applied to this problem.

Other Improvements

It is reported that in addition to the protein sequence, secondary structure formation depends on other factors. For example, it is reported that secondary structure tendencies depend also on local environment, solvent accessibility of residues, protein structural class, and even the organism from which the proteins are obtained. Based on such observations, some studies have shown that secondary structure prediction can be improved by addition of information about protein structural class, residue accessible surface area and also contact number information.

Sequence covariation methods rely on the existence of a data set composed of multiple homologous RNA sequences with related but dissimilar sequences. These methods analyze the covariation of individual base sites in evolution; maintenance at two widely separated sites of a pair of base-pairing nucleotides indicates the presence of a structurally required hydrogen bond between those positions. The general problem of pseudoknot prediction has been shown to be NP-complete.

Tertiary Structure

The practical role of protein structure prediction is now more important than ever. Massive amounts of protein sequence data are produced by modern large-scale DNA sequencing efforts such as the Human Genome Project. Despite community-wide efforts in structural genomics, the output of experimentally determined protein structures—typically by time-consuming and relatively expensive X-ray crystallography or NMR spectroscopy—is lagging far behind the output of protein sequences.

The protein structure prediction remains an extremely difficult and unresolved undertaking. The two main problems are calculation of protein free energy and finding the global minimum of this energy. A protein structure prediction method must explore the space of possible protein structures which is astronomically large. These problems can be partially bypassed in "comparative" or homology modeling and fold recognition methods, in which the search space is pruned by the assumption that the protein in question adopts a structure that is close to the experimentally determined structure of another homologous protein. On the other hand, the de novo or ab initio protein structure prediction methods must explicitly resolve these problems. The progress and challenges in protein structure prediction has been reviewed in Zhang 2008.

Ab Initio Protein Modelling

Ab initio- or de novo- protein modelling methods seek to build three-dimensional protein models "from scratch", i.e., based on physical principles rather than (directly) on previously solved structures. There are many possible procedures that either attempt to mimic protein folding or apply some stochastic method to search possible solutions (i.e., global optimization of a suitable energy function). These procedures tend to require vast computational resources, and have thus only been carried out for tiny proteins.

To predict protein structure de novo for larger proteins will require better algorithms and larger computational resources like those afforded by either powerful supercomputers (such as Blue Gene or MDGRAPE-3) or distributed computing (such as Folding@home, the Human Proteome Folding Project and Rosetta@Home). Although these computational barriers are vast, the potential benefits of structural genomics (by predicted or experimental methods) make ab initio structure prediction an active research field.

As an intermediate step towards predicted protein structures, contact map predictions have been proposed.

De Novo Protein Structure Prediction

In computational biology, de novo protein structure prediction is the task of estimating a protein's tertiary structure from its sequence alone. The problem is very difficult and has occupied leading scientists for decades. Research has focused in three areas: alternate lower-resolution representations of proteins, accurate energy functions, and efficient sampling methods. At present, the most successful methods have a reasonable probability of predicting the fold of a small protein domain within 5 angstroms.

De novo protein structure prediction methods attempt to predict tertiary structures from sequences based on general principles that govern protein folding energetics and/or statistical tendencies of conformational features that native structures acquire, without the use of explicit templates.

A general paradigm for de novo prediction involves sampling conformation space, guided by scoring functions and other sequence-dependent biases such that a large set of candidate ("decoy") structures are generated. Native-like conformations are then selected from these decoys using scoring functions as well as conformer clustering. High-resolution refinement is sometimes used as a final step to fine-tune native-like structures.

There are two major classes of scoring functions. Physics-based functions are based on mathematical models describing aspects of the known physics of molecular interaction. Knowledge-based functions are formed with statistical models capturing aspects of the properties of native protein conformations .

De novo methods tend to require vast computational resources, and have thus only been carried out for relatively small proteins. To predict protein structure de novo for larger proteins will require better algorithms and larger computational resources like those afforded by either powerful supercomputers (such as Blue Gene or MDGRAPE-3) or distributed computing projects (such as Folding@home, Rosetta@home, the Human Proteome Folding Project, or Nutritious Rice for the World). Although computational barriers are vast, the potential benefits of structural genomics (by predicted or experimental methods) make de novo structure prediction an active research field.

Comparative Protein Modelling

Comparative protein modelling uses previously solved structures as starting points, or templates. This is effective because it appears that although the number of actual proteins is vast, there is a limited set of tertiary structural motifs to which most proteins belong. It has been suggested that there are only around 2,000 distinct protein folds in nature, though there are many millions of different proteins.

Homology Modeling

Is based on the reasonable assumption that two homologous proteins will share very similar structures. Because a protein's fold is more evolutionarily conserved than its amino acid sequence, a target sequence can be modeled with reasonable accuracy on a very distantly related template, provided that the relationship between target and template can be discerned through sequence alignment. It has been suggested that the primary bottleneck in comparative modelling arises from difficulties in alignment rather than from errors in structure prediction given a known-good alignment. Unsurprisingly, homology modelling is most accurate when the target and template have similar sequences.

Protein Threading

scans the amino acid sequence of an unknown structure against a database of solved structures. In each case, a scoring function is used to assess the compatibility of the sequence to the structure, thus yielding possible three-dimensional models. This type of method is also known as 3D-1D fold recognition due to its compatibility analysis between three-dimensional structures and linear protein sequences. This method has also given rise to methods performing an inverse folding search by evaluating the compatibility of a given structure with a large database of sequences, thus predicting which sequences have the potential to produce a given fold.

Side-chain Geometry Prediction

Accurate packing of the amino acid side chains represents a separate problem in protein structure prediction. Methods that specifically address the problem of predicting side-chain geometry include dead-end elimination and the self-consistent mean field methods. The side chain conformations with low energy are usually determined on the rigid polypeptide backbone and using a set of discrete side chain conformations known as “rotamers.” The methods attempt to identify the set of rotamers that minimize the model’s overall energy.

These methods use rotamer libraries, which are collections of favorable conformations for each residue type in proteins. Rotamer libraries may contain information about the conformation, its frequency, and the standard deviations about mean dihedral angles, which can be used in sampling. Rotamer libraries are derived from structural bioinformatics or other statistical analysis of side-chain conformations in known experimental structures of proteins, such as by clustering the observed conformations for tetrahedral carbons near the staggered (60°, 180°, -60°) values.

Rotamer libraries can be backbone-independent, secondary-structure-dependent, or backbone-dependent. Backbone-independent rotamer libraries make no reference to backbone conformation, and are calculated from all available side chains of a certain type (for instance, the first example of a rotamer library, done by Ponder and Richards at Yale in 1987).

Secondary-structure-dependent libraries present different dihedral angles and/or rotamer frequencies for á-helix, â-sheet, or coil secondary structures. Backbone-dependent rotamer libraries present conformations and/or frequencies dependent on the local backbone conformation as defined by the backbone dihedral angles Õ and ø, regardless of secondary structure.

The modern versions of these libraries as used in most software are presented as multidimensional distributions of probability or frequency, where the peaks correspond to the dihedral-angle conformations considered as individual rotamers in the lists. Some versions are based on very carefully curated data and are used primarily for structure validation, while others emphasize relative frequencies in much larger data sets and are the form used primarily for structure prediction, such as the Dunbrack rotamer libraries. Side-chain packing methods are most useful for analyzing the protein's hydrophobic core, where side chains are more closely packed; they have more difficulty addressing the looser constraints and higher flexibility of surface residues, which often occupy multiple rotamer conformations rather than just one.

Prediction of Structural Classes

Statistical methods have been developed for predicting structural classes of proteins based on their amino acid composition, pseudo amino acid composition and functional domain composition.

Quaternary Structure

In the case of complexes of two or more proteins, where the structures of the proteins are known or can be predicted with high accuracy, protein–protein docking methods can be used to predict the structure of the complex. Information of the effect of mutations at specific sites on the affinity of the complex helps to understand the complex structure and to guide docking methods.

Protein–protein Interaction Prediction

Protein–protein interaction prediction is a field combining bioinformatics and structural biology in an attempt to identify and catalog physical interactions between pairs or groups of proteins. Understanding protein–protein interactions is important for the investigation of intracellular signaling pathways, modelling of protein complex structures and for gaining insights into various biochemical processes. Experimentally, physical interactions between pairs of proteins can be inferred from a variety of experimental techniques, including yeast two-hybrid systems, protein-fragment complementation assays (PCA), affinity purification/mass spectrometry, protein microarrays, fluorescence resonance energy transfer (FRET), and Microscale Thermophoresis (MST). Efforts to experimentally determine the interactome of numerous species are ongoing, and a number of computational methods for interaction prediction have been developed in recent years.

Methods

Proteins that interact are more likely to co-evolve, therefore, it is possible to make inferences about interactions between pairs of proteins based on their phylogenetic distances. It has also been observed in some cases that pairs of interacting proteins have fused orthologues in other organisms. In addition, a number of bound protein complexes have been structurally solved and can be used to identify the residues that mediate the interaction so that similar motifs can be located in other organisms.

Phylogenetic Profiling

Phylogenetic profiling finds pairs of protein families with similar patterns of presence or absence across large numbers of species. This method identifies pairs likely to act in the same biological process, but does not necessarily imply physical interaction. Prediction of co-evolved protein pairs based on similar phylogenetic trees

This method involves using a sequence search tool such as BLAST for finding homologues of a pair of proteins, then building multiple sequence alignments with alignment tools such as Clustal. From these multiple sequence alignments, phylogenetic distance matrices are calculated for each protein in the hypothesized interacting pair. If the matrices are sufficiently similar (as measured by their Pearson correlation coefficient) they are deemed likely to interact.

Identification of Homologous Interacting Pairs

This method consists of searching whether the two sequences have homologues which form a complex in a database of known structures of complexes.

The identification of the domains is done by sequence searches against domain databases such as Pfam using BLAST. If more than one complex of Pfam domains is identified, then the query sequences are aligned using a hidden Markov tool called HMMER to the closest identified homologues, whose structures are known. Then the alignments are analysed to check whether the contact residues of the known complex are conserved in the alignment.

Identification of Structural Patterns

This method builds a library of known protein–protein interfaces from the PDB, where the interfaces are defined as pairs of polypeptide fragments that are below a threshold slightly larger than the Van der Waals radius of the atoms involved. The sequences in the library are then clustered based on structural alignment and redundant sequences are eliminated. The residues that have a high (generally >50%) level of frequency for a given position are considered hotspots. This library is then used to identify potential interactions between pairs of targets, providing that they have a known structure (i.e. present in the PDB).

Bayesian Network Modelling

Bayesian methods integrate data from a wide variety of sources, including both experimental results and prior computational predictions, and use these features to assess the likelihood that a particular potential protein interaction is a true positive result. These methods are useful because experimental procedures, particularly the yeast two-hybrid experiments, are extremely noisy and produce many false positives, while the previously mentioned computational methods can only provide circumstantial evidence that a particular pair of proteins might interact.

3D Template-based Protein Complex Modelling

This method makes use of known protein complex structures to predict as well as structurally model interactions between query protein sequences. The prediction process generally starts by employing a sequence based method (e.g. Interolog) to search for protein complex structures that are homologous to the query sequences. These known complex structures are then used as templates to structurally model the interaction between query sequences. This method has the advantage of not only inferring protein interactions but also suggests models of how proteins interact structurally, which can provide some insights into the atomic level mechanism of that interaction. On the other hand, the ability for this method to makes a prediction is limited to a relatively small number of known protein complex structures.

Supervised Learning Problem

The problem of PPI prediction can be framed as a supervised learning problem. In this paradigm the known protein interactions supervise the estimation of a function that can predict whether an interaction exists or not between two proteins given data about the proteins (e.g., expression levels of each gene in different experimental conditions, location information, phylogenetic profile, etc.).

Relationship to Docking Methods

The field of protein–protein interaction prediction is closely related to the field of protein–protein docking, which attempts to use geometric and steric considerations to fit two proteins of known structure into a bound complex. This is a useful mode of inquiry in cases where both proteins in the pair have known structures and are known (or at least strongly suspected) to interact, but since so many proteins do not have experimentally determined structures, sequence-based interaction prediction methods are especially useful in conjunction with experimental studies of an organism's interactome.

Software

I-TASSER is the best server for protein structure prediction according to the 2006-2010 CASP experiments (CASP7, CASP8 and CASP9). RaptorX excels at aligning hard targets according to the 2010 CASP9 experiments. RaptorX generates the significantly better alignments for the hardest 50 CASP9 template-based modeling targets than other servers including those using consensus and refinement methods. The RaptorX server is available at server. MODELLER is a popular software tool for producing homology models using methodology derived from NMR spectroscopy data processing. Swiss Model provides an automated web server for basic homology modeling.

HHpred, bioinfo.pl and Robetta widely used servers for protein structure prediction. HHsearch is a free software package for protein threading and remote homology detection. SPARKSx is one of the top performing servers in the CASP focused on the remote fold recognition.

PEP-FOLD is a de novo approach aimed at predicting peptide structures from amino acid sequences, based on a HMM structural alphabet. Phyre and Phyre2 are amongst the top performing servers in the CASP international blind trials of structure prediction in homology modelling and remote fold recognition, and are designed with an emphasis on ease of use for non-experts.

RAPTOR (software) is a protein threading software that is based on integer programming. The basic algorithm for threading is described in and is fairly straightforward to implement.

QUARK is an on-line server suitable for ab initio protein structure modeling. Abalone is a Molecular Dynamics program for folding simulations with explicit or implicit water models.

TIP is a knowledgebase of STRUCTFAST models and precomputed similarity relationships between sequences, structures, and binding sites. Several distributed computing projects concerning protein structure prediction have also been implemented, such as the Folding@home, Rosetta@home, Human Proteome Folding Project, Predictor@home, and TANPAKU.

The Foldit program seeks to investigate the pattern-recognition and puzzle-solving abilities inherent to the human mind in order to create more successful computer protein structure prediction software.

Computational approaches provide a fast alternative route to antibody structure prediction. Recently developed antibody F_V region high resolution structure prediction algorithms, like RosettaAntibody, have been shown to generate high resolution homology models which have been used for successful docking. Reviews of software for structure prediction can be found at.

Automatic Structure Prediction Servers

CASP, which stands for Critical Assessment of Techniques for Protein Structure Prediction, is a community-wide experiment for protein structure prediction taking place every two years since 1994. CASP provides users and research groups with an opportunity to assess the quality of available methods and automatic servers for protein structure prediction. Official results for automatic structure prediction servers in the CASP7 benchmark (2006) are discussed by Battey et al.. Official CASP8 results are available for automatic servers and for human and server predictors. Unofficial results for automatic servers of the 2008 CASP8 benchmark are summarized on several lab websites and ranked according to slightly varying criteria: Zhang lab, Grishin lab, McGuffin lab, Baker lab, and Cheng lab.

Selection of Target Proteins

In order to ensure that no predictor can have prior information about a protein's structure that would put him/her at an advantage, it is important that the experiment is conducted in a double-blind fashion: Neither predictors nor the organizers and assessors know the structures of the target proteins at the time when predictions are made. Targets for structure prediction are either structures soon-to-be solved by X-ray crystallography or NMR spectroscopy, or structures that have just been solved (mainly by one of the structural genomics centers) and are kept on hold by the Protein Data Bank. If the given sequence is found to be related by common descent to a protein sequence of known structure (called a template), comparative protein modeling may be used to predict the tertiary structure. Templates can be found using sequence alignment methods such as BLAST or FASTA or protein threading methods, which are better in finding distantly related templates. Otherwise, de novo protein structure prediction must be applied, which is much less reliable but can sometimes yield models with the correct fold. Truly new folds are becoming quite rare among the targets, making that category smaller than desirable.

Evaluation

The primary method of evaluation is a comparison of the predicted model á-carbon positions with those in the target structure. The comparison is shown visually by cumulative plots of distances between pairs of equivalents á-carbon in the alignment of the model and the structure, such as shown in the figure (a perfect model would stay at zero all the way across), and is assigned a numerical score GDT-TS (Global Distance Test — Total Score) describing percentage of well-modeled residues in the model with respect to the target. Free modeling (template-free, or de novo) is also evaluated visually by the assessors, since the numerical scores do not work as well for finding loose resemblances in the most difficult cases. High-accuracy template-based predictions were evaluated in CASP7 by whether they worked for molecular-replacement phasing of the target crystal structure with successes followed up later, and by full-model (not just á-carbon) model quality and full-model match to the target in CASP8.

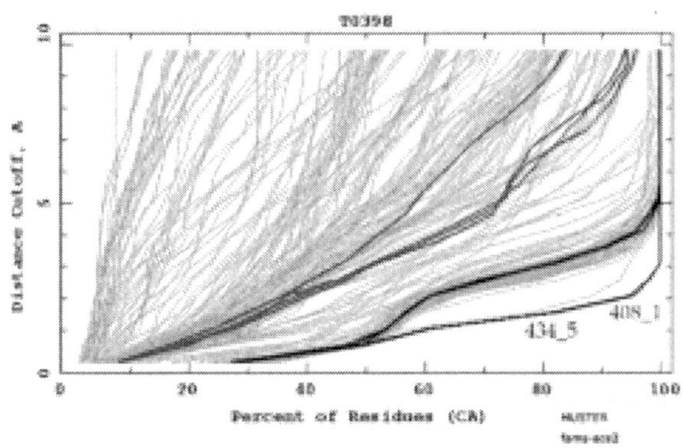

Figure: Cumulative plot of á-carbon accuracy, of all predicted models for target T0398 in CASP8, with the two best models labeled

Evaluation of the results is carried out in the following prediction categories:

- Tertiary structure prediction (all CASPs)
- Secondary structure prediction (dropped after CASP5)
- Prediction of structure complexes (CASP2 only; a separate experiment — CAPRI — carries on this subject)
- Residue-residue contact prediction (starting CASP4)
- Disordered regions prediction (starting CASP5)
- Domain boundary prediction (CASP6–CASP8)
- Function prediction (starting CASP6)
- Model quality assessment (starting CASP7)
- Model refinement (starting CASP7)
- high-accuracy template-based prediction (starting CASP7)

Tertiary structure prediction category was further subdivided into

- Homology modeling
- Fold recognition (also called protein threading; Note, this is incorrect as threading is a method)
- De novo structure prediction, now referred to as 'New Fold' as many methods apply evaluation, or scoring, functions that are biased by knowledge of native protein structures, such as an artificial neural network.

Starting with CASP7, categories have been redefined to reflect developments in methods. The 'Template based modeling' category includes all former comparative modeling, homologous fold based models and some analogous fold based models. The 'Template free modeling' category includes models of proteins with previously unseen folds and hard analogous fold based models.

The CASP results are published in special supplement issues of the scientific journal Proteins, all of which are accessible through the CASP website. A lead article in each of these supplements describes specifics of the experiment while a closing article evaluates progress in the field.

Energy Landscape of Protein Folding

The protein folding phenomenon was largely an experimental endeavor until the formulation of an energy landscape theory of proteins by Joseph Bryngelson and Peter Wolynes in the late 1980s and early 1990s. This approach introduced the principle of minimal frustration,. This principle says that nature has chosen amino acid sequences so that the folded state of the protein is very stable. Additionally, the undesired interactions between amino acids along the folding pathway are reduced making the acquisition of the folded state a very fast process. Even though nature has reduced the level of frustration in proteins, some degree of it remains up to now as can be observed in the presence of local minima in the energy landscape of proteins. A consequence of these evolutionarily selected sequences is that proteins are generally thought to have globally "funneled energy landscapes" (coined by José Onuchic) that are largely directed towards the native state. This "folding funnel" landscape allows the protein to fold to the native state through any of a large number of pathways and intermediates, rather than being restricted to a single mechanism. The theory is supported by both computational simulations of model proteins and experimental studies, and it has been used to improve methods for protein structure prediction and design. The description of protein folding by the leveling free-energy landscape is also consistent with the 2nd law of thermodynamics. Physically, thinking of landscapes in terms of visualizable potential or total energy surfaces simply with maxima, saddle points, minima and funnels, rather like geographic landscapes, is perhaps a little misleading. The relevant description is really a highly dimensional phase space in which manifolds might take a variety of more complicated topological forms.

Modeling of Protein Folding

De novo or ab initio techniques for computational protein structure prediction are related to, but strictly distinct from experimental studies of protein folding. Molecular Dynamics (MD) is an important tool for studying protein folding and dynamics in silico. First equilibrium folding simulations were done using implicit solvent model and Umbrella Sampling. Because of computational cost, ab initio MD folding simulations with explicit water are limited to peptides and very small proteins. MD simulations of larger proteins remain restricted to dynamics of the experimental structure or its high-temperature unfolding. In order to simulate long time folding processes (beyond about 1 microsecond), like folding of small-size proteins (about 50 residues) or larger, some approximations or simplifications in protein models need to be introduced. An approach using reduced protein representation (pseudo-atoms representing groups of atoms are defined) and statistical potential are useful in protein structure prediction and modeling of the folding pathways.

There are distributed computing projects which use idle CPU or GPU time of personal computers to solve problems such as protein folding or prediction of protein structure, one prominent example being the Folding@Home project. People can run these programs on their computer or PlayStation 3 to support them.

Protein Transport

Many proteins are destined for other parts of the cell than the cytosol and a wide range of signalling sequences are used to direct proteins to where they are supposed to be. In prokaryotes this is normally a simple process due to limited compartmentalisation of the cell. However in eukaryotes there is a great variety of different targeting processes to ensure the protein arrives at the correct organelle.

Not all proteins remain within the cell and many are exported, for example digestive enzymes, hormones and extracellular matrix proteins. In eukaryotes the export pathway is well developed and the main mechanism for the export of these proteins is translocation to the endoplasmic reticulum, followed by transport via the Golgi apparatus.

Design

DNA nanostructures must be rationally designed so that the individual nucleic acid strands will assemble into the desired structures. The design process of such nanostructures usually begins with the specification of a desired target structure and/or functionality. Then, the overall secondary structure of the target molecule is designed, meaning the arrangement of nucleic acid strands within the structure, and which portions of those strands should be bound to each other. The last step is the primary structure design, the specification of the actual base sequences of each nucleic acid strand.

Structural Design

The first step in designing a nucleic acid nanostructure is to decide how a given structure should be represented by a specific arrangement of nucleic acid strands. This design step thus determines the secondary structure, or the series of base pairs which hold the individual strands together in the desired shape. There are several approaches which have been demonstrated:

- Tile-based structures. This approach breaks the target structure into smaller units with strong binding between the strands contained in each unit, and relatively weaker interactions between the units. It is often used to make periodic lattices, but can also be used to implement algorithmic self-assembly, making them one platform for DNA computing.
- Folding structures. An alternative to the tile-based approach, folding approaches make the nanostructure out of a single long strand. This long strand can either have a designed sequence which folds due to its interactions with itself, or it can be folded into the desired shape by using shorter, "staple" strands. This latter method is called DNA origami, which allows the creation of two- and three-dimensional shapes at the nanoscale using DNA.
- Kinetic assembly. Recently, there has been interest in controlling the kinetics of DNA self-assembly, so that transient dynamics can also be programmed into the assembly. Such a method also has the advantage of proceeding isothermally and thus not requiring a thermal annealing step required by solely thermodynamic approaches.

Sequence Design

After any of the above approaches are used to design the secondary structure of a target molecule, an actual sequence of nucleotides must be devised which will form into the desired structure. Nucleic acid design is the process of assigning a specific nucleic acid base sequence to each strand so that they will associate into a desired conformation. Nucleic acid design is central to the field of DNA nanotechnology. Most methods seek to designing sequences so that the target structure is a thermodynamic minimum, and mis-assembled structures have higher energies and are thus disfavored.

This is done either through heuristic methods such as sequence symmetry minimization and coding theory based approaches, or by explicitly using a full nearest-neighbor thermodynamic model. Geometric models are also used to examine tertiary structure of the nanostructures and ensure that the complexes are not overly strained.

Nucleic acid design has similar goals to protein design: in both, the sequence of monomers is designed to favor the desired folded or associated structure and to disfavor alternate structures. Nucleic acid design has the advantage of being a much computationally simpler problem, since the simplicity of Watson-Crick base pairing rules leads to simple heuristic methods which yield experimentally robust designs. However, nucleic acid structures are less versatile than proteins in their functionality.

Structural DNA Nanotechnology

Structural DNA nanotechnology, sometimes abbreviated as SDN, focuses on synthesizing and characterizing nucleic acid complexes and materials with various nanoscale structures. Structural DNA nanotechnology is largely based on the fact that the three-dimensional structure of DNA—the nucleic acid double helix— has a robust, defined geometry which makes it possible to predict and design the structures of more complex DNA molecules. Many such structures have been created, including two- and three-dimensional structures; and periodic, aperiodic, and discrete structures.

Periodic Lattices

Smaller nucleic acid assemblies can be equipped with sticky ends in order to combine them into a two-dimensional periodic lattice. The earliest example of this was the array of DX, or double-crossover, molecules. Each DX molecule can be designed with four sticky ends, one at each end of the two double-helical domains, and these sticky ends can be designed with sequences that cause the DX units to combine into a specific tessellated pattern. They thus form extended flat sheets which are essentially rigid two-dimensional crystals of DNA.

Two-dimensional arrays have been made out of other motifs as well, including the Holliday junction rhombus array as well as various DX-based arrays making use of a double-cohesion scheme.

Creating three-dimensional lattices out of DNA was the earliest goal of DNA nanotechnology, but proved to be one of the most difficult to realize. Success in constructing three-dimensional DNA lattices was finally reported in 2009 using a motif based on the concept of tensegrity, a balance between tension and compression forces.

Nanotubes

In addition to flat sheets, DX arrays have been made to form hollow nanotubes of 4–20 nm diameter. These DNA nanotubes are somewhat similar in size and shape to carbon nanotubes, but the carbon nanotubes are stronger and better conductors, whereas the DNA nanotubes are more easily modified and connected to other structures. There have been multiple schemes for constructing DNA nanotubes, one of which uses the inherent curvature of DX tiles to form a DX lattice to curl around itself and close into a tube. An alternative design uses single-stranded "tiles" for which the rigidity of the tube is an emergent property. This method also has the benefit of being able to determine the circumference of the nanotube in a simple, modular fashion.

Polyhedra

A number of three-dimensional DNA molecules have been made which have the connectivity of a polyhedron such as an octahedron or cube. In other words, the DNA duplexes trace the edges of a polyhedron with a DNA junction at each vertex. The earliest demonstrations of DNA polyhedra involved multiple ligations and solid-phase synthesis steps to create catenated polyhedra. More recent work has yielded polyhedra whose synthesis is much easier. These include a DNA octahedron made from a long single strand designed to fold into the correct conformation, as well as a tetrahedron which can be produced from four DNA strands in a single step, pictured at the top of this article.

Arbitrary Shapes

Nanostructures of arbitrary shapes are usually made using the DNA origami method. DNA origami makes use of a long natural virus strand as a "scaffold" strand, and computationally designs shorter "staple" strands which bind to portions of the scaffold strand and force it to fold into the desired shape. This method has the advantage of being easy to design, as the base sequence is predetermined by the scaffold strand sequence, and it also does not require high strand purity and accurate stoichiometry, as most other DNA nanotechnology methods do. DNA origami was first demonstrated for two-dimensional shapes; demonstrated designs included the smiley face and a coarse map of North America. This was later extended to solid three-dimensional shapes. In addition, structures have been constructed with two-dimensional faces which fold into an overall three-dimensional shape, akin to a cardboard box. These can be programmed to open and release their cargo in response to a stimulus, making them potentially useful as programmable molecular cages.

Functional Nucleic acid Nanostructures

DNA nanotechnology focuses on creating molecules with designed functionalities as well as structures. These include both dynamic functionality within the nucleic acid structure itself, for example with computation and mechanical motion, as well as by including other components such as small molecules or nanoparticles which have their own functionalities. Many classes of functional systems have been demonstrated.

Nanoarchitecture

The idea of using DNA arrays to template the assembly of other functional molecules was first suggested by Nadrian Seeman in 1987, but only recently has progress been made in reducing these kinds of schemes to practice. In 2006, researchers covalently attached gold nanoparticles to a DX-based tile and showed that self-assembly of the DNA structures also assembled the nanoparticles hosted on them. Also that year, Dwyer and LaBean demonstrated the letters "D" "N" and "A" created on a 4x4 DX array using streptavidin, and a hierarchical assembly based on this approach was also demonstrated that scales to larger arrays (8X8 and 8.96 MD). A non-covalent hosting scheme was shown in 2007, using Dervan polyamides on a DX array to arrange streptavidin proteins on specific kinds of tiles on the DNA array.

There has also been interest in using DNA nanotechnology to assemble molecular electronics devices. To this end, DNA has been used to assemble single walled carbon nanotubes into field-effect transistors.

Algorithmic Self-assembly

DNA nanotechnology has been applied to the related field of DNA computing. The DX tiles can have their sticky end sequences chosen so that they act as Wang tiles, allowing them to perform computation. A DX array has been demonstrated whose assembly encodes an XOR operation; this allows the DNA array to implement a cellular automaton which generates a fractal called the Sierpinski gasket. Another system has the function of a binary counter, displaying a representation of increasing binary numbers as it grows. These results show that computation can be incorporated into the assembly of DNA arrays, increasing its scope beyond simple periodic arrays.

Figure: The Sierpinski gasket fractal.

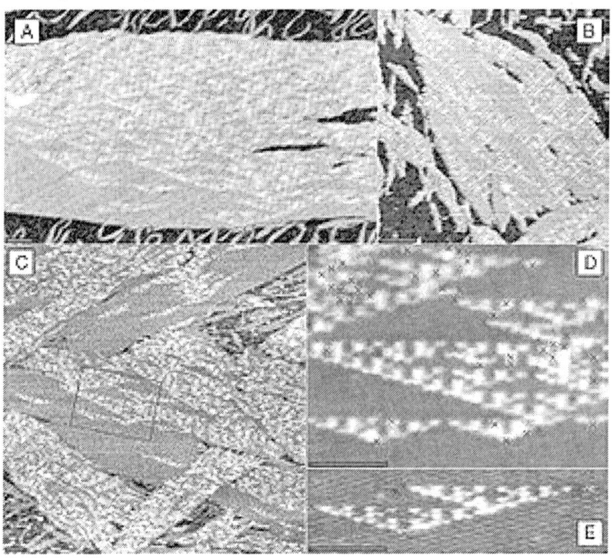

Figure: DNA arrays that display a representation of the Sierpinski gasket on their surfaces.

Note that DNA computing overlaps with, but is distinct from, DNA nanotechnology. The latter uses the specificity of Watson-Crick basepairing to make novel structures out of DNA. These structures can be used for DNA computing, but they do not have to be. Additionally, DNA computing can be realized without using the types of molecules made possible by DNA nanotechnology.

Nanomechanical Devices

DNA complexes have been made which change their conformation upon some stimulus. These are intended to have applications in nanorobotics. DNA machines have also been made which show a twisting motion. The first such device made use of the transition between the B-DNA and Z-DNA forms to respond to a change in buffer conditions. This reliance on buffer conditions, however, caused all devices to change state at the same time. A subsequent system, called "molecular tweezers," changes from an open to a closed state based upon the presence of control strands, allowing multiple devices to be individually operated in solution. This was followed up by another system which relies on the presence of control strands to switch from a paranemic-crossover (PX) conformation to a double-junction (JX2) conformation.

Nucleic acid nanomachines have been made which exhibit directional motion along a linear track, called DNA walkers. A large number of schemes have been demonstrated. One strategy is to control the motion of the walker along the track using control strands which need to be manually added in sequence. Another approach is to make use of restriction enzymes or deoxyribozymes to cleave the strands and cause the walker to move forward, which has the advantage of running autonomously. A later system extended the concept of DNA walkers to walk upon a two-dimensional surface rather than a linear track, and demonstrated the ability to selectively pick up and move molecular cargo. Additionally, a linear walker has been demonstrated which performs DNA-templated synthesis as the walker advances along the track, allowing autonomous multistep chemical synthesis directed by the walker.

Applications

DNA nanotechnology provides one of the only ways to form designed, complex structures with precise control over nanoscale features. The field is beginning to see application to solve basic science problems in structural biology and biophysics. One such application, the earliest envisaged for the field, is in crystallography, where molecules which are hard to crystallize by themselves could be arranged and oriented within a three-dimensional nucleic acid lattice, thus allowing determination of their structure. DNA origami rods have also been used to replace liquid crystals in residual dipolar coupling experiments in protein NMR spectroscopy; using DNA origami is advantageous because, unlike liquid crystals, they are tolerant of the detergents needed to suspend membrane proteins in solution. It has been demonstrated that DNA walkers can be used as nanoscale assembly lines to move nanoparticles and direct chemical synthesis. Furthermore, DNA origami structures have aided in the biophysical studies of enzyme function and protein folding.

DNA nanotechnology is also moving towards potential real-world applications. It has been suggested that the ability of nucleic acid arrays to arrange other molecules has potential applications in molecular scale electronics, with the assembly of a nucleic acid lattice templating the assembly of molecular electronic elements such as molecular wires. DNA nanotechnology has also been called a form of programmable matter because of the coupling of computation to its material properties.

There are also potential applications for DNA nanotechnology in nanomedicine, making use of its ability to perform computation in a biocompatible format to make "smart drugs" for targeted drug delivery. One such system being investigated uses a hollow DNA box containing proteins which induce apoptosis, or cell death, which will only open when in proximity to a cancer cell.

Materials and Methods

The sequences of the individual DNA strands which make up the target structure are designed computationally, using molecular modeling and thermodynamic modeling software. Once the sequences have been designed, the nucleic acid molecules themselves are synthesized through standard oligonucleotide synthesis methods. This process is usually automated by using a machine called an oligonucleotide synthesizer, and nucleic acids of custom sequence are commercially available from many vendors. For methods which require pure strands of known concentration, the nucleic acid strands can be purified using denaturing gel electrophoresis, and concentrations are determined by one of several nucleic acid quantitation methods using ultraviolet absorbance spectroscopy.

The fully formed target structures are usually characterized by native gel electrophoresis, which provides information about the size and shape of DNA molecules. An electrophoretic mobility shift assay can indicate whether a stucture incorporates all the individual desired strands. Fluorescent labeling and Förster resonance energy transfer (FRET) are also used to characterize the structure of the molecules.

Nucleic acid structures can be directly imaged by atomic force microscopy, which is well-suited to extended two-dimensional structures, but is less useful for discrete three-dimensional structures due to the interaction of the microscope tip with the fragile nucleic acid structure.

For these latter structures transmission electron microscopy and cryo-electron microscopy are important methods. Extended three-dimensional lattices are analyzed by X-ray crystallography.

History

The concept of DNA nanotechnology was invented by Nadrian Seeman in the early 1980s. Seeman was originally concerned with using a three-dimensional DNA lattice to orient target molecules, which would simplify their crystallographic study by eliminating the difficult process of obtaining pure crystals. This idea had reportedly come to him in fall 1980, after realizing the similarity between the M. C. Escher woodcut Depth and an array of DNA six-arm junctions.

To this end, Seeman's laboratory in 1991 published the synthesis of a cube made of DNA, the first three-dimensional nanoscale object, for which he received the 1995 Feynman Prize in Nanotechnology, which was followed by a DNA truncated octahedron. However, it soon became clear that these molecules, polygonal shapes with flexible junctions as their vertices, were not rigid enough to form extended three-dimensional lattices. Seeman developed the more rigid double-crossover (DX) motif, and in collaboration with Erik Winfree, in 1998 published the creation of two-dimensional lattices of DX tiles. The synthesis of a three-dimensional lattice was finally published by Seeman in 2009, nearly thirty years after he had set out to do so.

These tile-based structures had the advantage that they provided the capability to implement DNA computing, which was demonstrated by Winfree and Paul Rothemund in their 2004 paper on the algorithmic self-assembly of a Sierpinski gasket structure, and for which they shared the 2006 Feynman Prize in Nanotechnology. Winfree's key insight was that the DX tiles could be used as Wang tiles, meaning that their assembly was capable of performing computation.

The first DNA nanomachine—a motif which changes its structure in response to an input—was demonstrated in 1999 by Seeman. An improved system was demonstrated by Bernard Yurke the following year, which was also the first nucleic acid device to make use of toehold-mediated strand displacement. The next advance was to translate this into mechanical motion, and in 2004 and 2005, a number of DNA walker systems were demonstrated by the groups of Seeman, Niles Pierce, Andrew Turberfield, and Chengde Mao.

In 2006, Rothemund first demonstrated the new DNA origami technique for easily and robustly creating folded DNA molecules of any shape. Rothemund had conceived of this method as being conceptually intermediate between Seeman's DX lattices, which used many short strands, and William Shih's DNA octahedron, which consisted mostly of one very long strand; Rothemund's DNA origami contains a long strand whose folding is assisted by a number of short strands. This method allowed the creation of much larger structures than were previously possible, which are less technically demanding to design and synthesize. DNA origami was the cover story of Nature on March 15, 2006. Rothemund's research demonstrating two-dimensional DNA origami structures was followed by the demonstration of solid three-dimensional DNA origami by Shih in 2009, while the labs of Jørgen Kjems and Hao Yan demonstrated hollow three-dimensional structures made out of two-dimensional faces.

DNA nanotechnology as a field was initially met with some skepticism due to the unusual non-biological use of nucleic acids as a material for building structures and doing computation, as well as the preponderance of proof of principle experiments which extended the capabilities of the field but were far from actual applications. Seeman's seminal 1991 paper on the synthesis of the DNA cube was rejected by the journal Science after one reviewer praised its originality while another criticized it for its lack of biological relevance.

By the early 2010s, however, the field was considered to have increased its capabilities to the point that applications for basic science research were beginning to be realized, and practical applications in medicine and other fields were beginning to be considered feasible. In addition, the field had grown from very few active laboratories in 2001, to at least 60 in 2010, which increased the talent pool and thus the number of advancements in the field during that decade.

History and Anthropology

Because DNA collects mutations over time, which are then inherited, it contains historical information, and, by comparing DNA sequences, geneticists can infer the evolutionary history of organisms, their phylogeny. This field of phylogenetics is a powerful tool in evolutionary biology. If DNA sequences within a species are compared, population geneticists can learn the history of particular populations. This can be used in studies ranging from ecological genetics to anthropology; For example, DNA evidence is being used to try to identify the Ten Lost Tribes of Israel.

DNA has also been used to look at modern family relationships, such as establishing family relationships between the descendants of Sally Hemings and Thomas Jefferson. This usage is closely related to the use of

DNA in criminal investigations detailed above. Indeed, some criminal investigations have been solved when DNA from crime scenes has matched relatives of the guilty individual.

History of DNA Research

DNA was first isolated by the Swiss physician Friedrich Miescher who, in 1869, discovered a microscopic substance in the pus of discarded surgical bandages.

As it resided in the nuclei of cells, he called it "nuclein". In 1878, Albrecht Kossel isolated the non-protein component of "nuclein", nucleic acid, and later isolated its five primary nucleobases. In 1919, Phoebus Levene identified the base, sugar and phosphate nucleotide unit. Levene suggested that DNA consisted of a string of nucleotide units linked together through the phosphate groups. However, Levene thought the chain was short and the bases repeated in a fixed order. In 1937 William Astbury produced the first X-ray diffraction patterns that showed that DNA had a regular structure.

In 1927 Nikolai Koltsov proposed that inherited traits would be inherited via a "giant hereditary molecule" which would be made up of "two mirror strands that would replicate in a semi-conservative fashion using each strand as a template". In 1928, Frederick Griffith discovered that traits of the "smooth" form of the Pneumococcus could be transferred to the "rough" form of the same bacteria by mixing killed "smooth" bacteria with the live "rough" form. This system provided the first clear suggestion that DNA carries genetic information—the Avery–MacLeod–McCarty experiment—when Oswald Avery, along with coworkers Colin MacLeod and Maclyn McCarty, identified DNA as the transforming principle in 1943. DNA's role in heredity was confirmed in 1952, when Alfred Hershey and Martha Chase in the Hershey–Chase experiment showed that DNA is the genetic material of the T2 phage.

In 1953, James D. Watson and Francis Crick suggested what is now accepted as the first correct double-helix model of DNA structure in the journal Nature. Their double-helix, molecular model of DNA was then based on a single X-ray diffraction image (labeled as "Photo 51") taken by Rosalind Franklin and Raymond Gosling in May 1952, as well as the information that the DNA bases are paired — also obtained through private communications from Erwin Chargaff in the previous years. Chargaff's rules played a very important role in establishing double-helix configurations for B-DNA as well as A-DNA. Experimental evidence supporting the Watson and Crick model were published in a series of five articles in the same issue of Nature. Of these, Franklin and Gosling's paper was the first publication of their own X-ray diffraction data and original analysis method that partially supported the Watson and Crick model; this issue also contained an article on DNA structure by Maurice Wilkins and two of his colleagues, whose analysis and in vivo B-DNA X-ray patterns also supported the presence in vivo of the double-helical DNA configurations as proposed by Crick and Watson for their double-helix molecular model of DNA in the previous two pages of Nature.

In 1962, after Franklin's death, Watson, Crick, and Wilkins jointly received the Nobel Prize in Physiology or Medicine. However, Nobel rules of the time allowed only living recipients, but a vigorous debate continues on who should receive credit for the discovery.

In an influential presentation in 1957, Crick laid out the central dogma of molecular biology, which foretold the relationship between DNA, RNA, and proteins, and articulated the "adaptor hypothesis". Final confirmation of the replication mechanism that was implied by the double-helical structure followed in 1958 through the Meselson–Stahl experiment. Further work by Crick and coworkers showed that the genetic code was based on non-overlapping triplets of bases, called codons, allowing Har Gobind Khorana, Robert W. Holley and Marshall Warren Nirenberg to decipher the genetic code. These findings represent the birth of molecular biology.

Crystallography

Crystallography is the experimental science of the arrangement of atoms in solids. The word "crystallography" derives from the Greek words crystallon = cold drop / frozen drop, with its meaning extending to all solids with some degree of transparency, and grapho = write.

Before the development of X-ray diffraction crystallography, the study of crystals was based on their geometry. This involves measuring the angles of crystal faces relative to theoretical reference axes (crystallographic axes), and establishing the symmetry of the crystal in question. The former is carried out using a goniometer. The position in 3D space of each crystal face is plotted on a stereographic net, e.g. Wulff net or Lambert net. In fact, the pole to each face is plotted on the net. Each point is labelled with its Miller index. The final plot allows the symmetry of the crystal to be established.

Crystallographic methods now depend on the analysis of the diffraction patterns of a sample targeted by a beam of some type. Although X-rays are most commonly used, the beam is not always electromagnetic radiation. For some purposes electrons or neutrons are used. This is facilitated by the wave properties of the particles. Crystallographers often explicitly state the type of illumination used when referring to a method, as with the terms X-ray diffraction, neutron diffraction and electron diffraction.

These three types of radiation interact with the specimen in different ways. X-rays interact with the spatial distribution of the valence electrons, while electrons are charged particles and therefore feel the total charge distribution of both the atomic nuclei and the surrounding electrons. Neutrons are scattered by the atomic nuclei through the strong nuclear forces, but in addition, the magnetic moment of neutrons is non-zero.

They are therefore also scattered by magnetic fields. When neutrons are scattered from hydrogen-containing materials, they produce diffraction patterns with high noise levels. However, the material can sometimes be treated to substitute hydrogen for deuterium. Because of these different forms of interaction, the three types of radiation are suitable for different crystallographic studies.

Theory

Generally, an image of a small object is made using a lens to focus the illuminating radiation, as is done with the rays of the visible spectrum in light microscopy. However, the wavelength of visible light (about 4000 to 7000 angstroms) is three orders of magnitude longer than the length of typical atomic bonds and atoms themselves (about 1 to 2 angstroms).

Therefore, obtaining information about the spatial arrangement of atoms requires the use of radiation with shorter wavelengths, such as X-rays or neutrons beam. Employing shorter wavelengths implied abandoning microscopy and true imaging, however, because there exists no material from which a lens capable of focusing this type of radiation can be created. (That said, scientists have had some success focusing X-rays with microscopic Fresnel zone plates made from gold, and by critical-angle reflection inside long tapered capillaries.)

Diffracted x-ray or neutrons beams cannot be focused to produce images, so the sample structure must be reconstructed from the diffraction pattern. Sharp features in the diffraction pattern arise from periodic, repeating structure in the sample, which are often very strong due to coherent reflection of many photons from many regularly spaced instances of similar structure, while non-periodic components of the structure result in diffuse (and usually weak) diffraction features. Said more simply, areas with a higher density and repetition of atom order tend to reflect more light toward one point in space when compared to those areas with fewer atoms and less repetition.

Because of their highly ordered and repetitive structure, crystals give diffraction patterns of sharp Bragg reflection spots, and are ideal for analyzing the structure of solids.

Notation

- Coordinates in square brackets such as [100] denote a direction vector (in real space).
- Coordinates in angle brackets or chevrons such as <100> denote a family of directions which are related by symmetry operations. In the cubic crystal system for example, <100> would mean [100], [010], [001] or the negative of any of those directions.
- Miller indices in parentheses such as (100) denote a plane of the crystal structure, and regular repetitions of that plane with a particular spacing. In the cubic system, the normal to the (hkl) plane is the direction [hkl], but in lower-symmetry cases, the normal to (hkl) is not parallel to [hkl].
- Indices in curly brackets or braces such as {100} denote a family of planes and their normals which are equivalent in cubic materials due to symmetry operations, much the way angle brackets denote a family of directions. In non-cubic materials, <hkl> is not necessarily perpendicular to {hkl}.

Technique

Some materials studied using crystallography, proteins for example, do not occur naturally as crystals. Typically, such molecules are placed in solution and allowed to crystallize over days, weeks, or months through vapor diffusion. A drop of solution containing the molecule, buffer, and precipitants is sealed in a container with a reservoir containing a hygroscopic solution. Water in the drop diffuses to the reservoir, slowly increasing the concentration and allowing a crystal to form. If the concentration were to rise more quickly, the molecule would simply precipitate out of solution, resulting in disorderly granules rather than an orderly and hence usable crystal.

Once a crystal is obtained, data can be collected using a beam of radiation. Although many universities that engage in crystallographic research have their own X-ray producing equipment, synchrotrons are often used as X-ray sources, because of the purer and more complete patterns such sources can generate. Synchrotron sources also have a much higher intensity of X-ray beams, so data collection takes a fraction of the time normally necessary at weaker sources. Complementary neutron crystallography techniques are used to enhance hydrogen atoms positions. Such techniques are available in Neutron facilities.

Producing an image from a diffraction pattern requires sophisticated mathematics and often an iterative process of modelling and refinement. In this process, the mathematically predicted diffraction patterns of an hypothesized or "model" structure are compared to the actual pattern generated by the crystalline sample. Ideally, researchers make several initial guesses, which through refinement all converge on the same answer. Models are refined until their predicted patterns match to as great a degree as can be achieved without radical revision of the model. This is a painstaking process, made much easier today by computers.

The mathematical methods for the analysis of diffraction data only apply to patterns, which in turn result only when waves diffract from orderly arrays. Hence crystallography applies for the most part only to crystals, or to molecules which can be coaxed to crystallize for the sake of measurement. In spite of this, a certain amount of molecular information can be deduced from the patterns that are generated by fibers and powders, which while not as perfect as a solid crystal, may exhibit a degree of order.

This level of order can be sufficient to deduce the structure of simple molecules, or to determine the coarse features of more complicated molecules. For example, the double-helical structure of DNA was deduced from an X-ray diffraction pattern that had been generated by a fibrous sample.

Crystallography in Materials Engineering

Crystallography is a tool that is often employed by materials scientists. In single crystals, the effects of the crystalline arrangement of atoms is often easy to see macroscopically, because the natural shapes of crystals reflect the atomic structure. In addition, physical properties are often controlled by crystalline defects. The understanding of crystal structures is an important prerequisite for understanding crystallographic defects. Mostly, materials do not occur in a single crystalline, but poly-crystalline form, such that the powder diffraction method plays a most important role in structural determination. A number of other physical properties are linked to crystallography. For example, the minerals in clay form small, flat, platelike structures.

Clay can be easily deformed because the platelike particles can slip along each other in the plane of the plates, yet remain strongly connected in the direction perpendicular to the plates. Such mechanisms can be studied by crystallographic texture measurements.

In another example, iron transforms from a body-centered cubic (bcc) structure to a face-centered cubic (fcc) structure called austenite when it is heated. The fcc structure is a close-packed structure, and the bcc structure is not, which explains why the volume of the iron decreases when this transformation occurs. Crystallography is useful in phase identification. When performing any process on a material, it may be desired to find out what compounds and what phases are present in the material. Each phase has a characteristic arrangement of atoms. Techniques like X-ray or neutron diffraction can be used to identify which patterns are present in the material, and thus which compounds are present. Crystallography covers the enumeration of the symmetry patterns which can be formed by atoms in a crystal and for this reason has a relation to group theory and geometry.

Biology

X-ray crystallography is the primary method for determining the molecular conformations of biological macromolecules, particularly protein and nucleic acids such as DNA and RNA. In fact, the double-helical structure of DNA was deduced from crystallographic data. The first crystal structure of a macromolecule was solved in 1958 A three-dimensional model of the myoglobin molecule obtained by X-ray analysis. The Protein Data Bank (PDB) is a freely accessible repository for the structures of proteins and other biological macromolecules. Computer programs like RasMol or Pymol can be used to visualize biological molecular structures. Neutron crystallography is often used to help refine structures obtained by x-ray methods or to solve a specific bond; the methods are often viewed as complementary, as x-rays are sensitive to electron positions and scatter most strongly off heavy atoms, while neutrons are sensitive to nucleus positions and scatter strongly off many light isotopes, including hydrogen and deuterium. Electron crystallography has been used to determine some protein structures, most notably membrane proteins and viral capsids.

DNA Error

The Singularity is Near

The Singularity Is Near: When Humans Transcend Biology is a 2005 update of Raymond Kurzweil's 1999 book, The Age of Spiritual

Machines and his 1990 book The Age of Intelligent Machines. In it, as in the two previous versions, Kurzweil attempts to give a glimpse of what awaits us in the near future. He proposes a coming technological singularity, and how we would thus be able to augment our bodies and minds with technology. He describes the singularity as resulting from a combination of three important technologies of the 21st century: genetics, nanotechnology, and robotics (including artificial intelligence).

Four central postulates of the book are as follows:

1. A technological-evolutionary point known as "the singularity" exists as an achievable goal for humanity.
2. Through a law of accelerating returns, technology is progressing toward the singularity at an exponential rate.
3. The functionality of the human brain is quantifiable in terms of technology that we can build in the near future.
4. Medical advancements make it possible for a significant number of his generation (Baby Boomers) to live long enough for the exponential growth of technology to intersect and surpass the processing of the human brain.

Kurzweil's speculative reasoning and selective use of growth indicators has been heavily debated and challenged. In response to this, in the last chapter he gives responses to some of the criticisms he has received.

Postulates

Acceptance and striving for the idea of living forever

Kurzweil asserts that a technological-evolutionary jump known as "the singularity" will be recognized as an achievable goal by humanity. Kurzweil acknowledges that belief in the singularity promotes the paradigm shift necessary for its advancement. By promoting the "truth" of its coming through predictions that seem remarkable at the time but inevitable after the fact (a global computer network, a computer beating the chess champion, etc.), Kurzweil's popular series of books reinforces the belief that a singularity is unavoidable.

The Law of Accelerating Returns

Kurzweil asserts in his Law of Accelerating Returns that technology is progressing toward the Singularity at an exponential rate, relying almost entirely on empirical data. He expands on Moore's Law with models showing that not only the return, but the rate of return is increasing exponentially.

An Objective Measurement of Cerebral Processing Power

Kurzweil asserts that the functionality of the brain is quantifiable in terms of technology that we can build in the near future. Kurzweil's earlier books showed cerebral processing power as primarily the number of computations in a square inch multiplied by the area of the brain. In this update, however, he acknowledges the possibility of Penrose-Hameroff Microtubule quantum processing (Orch-OR) and states that if his calculations of the processing capability of the brain are off by a factor of a billion, the double-exponential growth of technology will still catch up to it twenty-four years after his original projections. The Orch-OR theory is generally discredited among neuroscientists.

In a rebuttal paper, Hameroff asserts that the quantum processing power required for consciousness is at an order of magnitude greater than can be expressed through conventional systems of processing measurement. This argument seems to ignore Kurzweil's premise that accelerating returns in development of present technologies could, in a nominal period, overcome such a barrier. Additionally, other technologies could emerge which greatly lower the time required to reach the Singularity. A notable example would be more advanced forms of quantum computing capable of full neural emulation, which on a large enough scale, could function equivalently to the biological human brain.

Sufficient Medical Advancements

Kurzweil asserts that medical advancements will keep his generation alive long enough for the exponential growth of technology to intersect and surpass the processing of the human brain. Kurzweil explains how nanobots will eventually be able to repair and replace any part of the body that wears out, but relies on other methods of medical technology to prolong our lives long enough to reach the singularity. The usefulness of this medical postulate then becomes a function of how long it will take to reach the singularity, something that has been thrown into question due to the possibility of quantum brain processing explored in many recent books by scientists, such as Roger Penrose's The Emperor's New Mind.

DNA Errors

According to Ray Kurzweil, DNA errors (in the form of either pathogens, viruses, and/or cancer cells) are the causes of most major illnesses. These illnesses range from cancer to Alzheimer's disease and even Down syndrome. 130 different nanotechnology-related devices were invented in 2006 that could possibly eliminate the DNA errors that cause cancer someday. A majority of the statements about nanobots eliminating DNA errors and pathogens were emphasized frequently in Kurzweil's futurist book The Singularity Is Near; published in 2005 by an American publishing company.

The only permanent way to get rid of the DNA errors (that are a part of cancer cells and/or the various pathogens found in other illnesses) is by injecting nanobots (ideally multifunctional nanoparticles) into the patient's body. The high levels of SP2 protein that are found in cancer cells would be neutralized. The same nanobots that would lower the levels of SP2 protein inside the patient's stem cells could also repair the long-term damage to the person's DNA strand so that his/her descendants are less likely to acquire cancer and/other diseases in the first place.

All methods of death that are not caused by accidents are caused by a long line of DNA errors that are never corrected over a very long period of time. However, this method may not be possible or even affordable until the year 2025 (the Kanzius RF Therapy is scheduled to begin testing in 2012 but will be extremely difficult to acquire for many years). All research about curing cancer with nanorobotics has been currently restricted to small animals like mice. Thanks to the cost-performance ratio, nanomedicine is becoming more affordable for the average person year by year. It may become possible to have extremely effective and extremely inexpensive treatments for AIDS that will eliminate the DNA errors that are causing the disease through nanomedicine.

Even with the developed countries of the world competing against each other to be the first to create a sustainable nanorobotics program that can cure cancers and other diseases caused by DNA errors, government oversight (in addition to further oversight by the United Nations) will allow the DNA error-eliminating procedures to become increasingly affordable to everyone.

Organization

- Prologue: The Power of Ideas
- Chapter 1: The Six Epochs
- Chapter 2: A Theory of Technology Evolution
- Chapter 3: Achieving the Computational Capacity of the Human Brain
- Chapter 4: Achieving the Software of Human Intelligence
- Chapter 5: GNR (genetics, nanotech, robots)
- Chapter 6: The Impact

- Chapter 7: Ich bin ein Singularitarian
- Chapter 8: The Deeply Intertwined Promise and Peril of GNR
- Chapter 9: Response to critics
- Epilogue

Kurzweil first defines the Singularity as a point in the future when technological advances begin to happen so rapidly that normal humans cannot keep pace, and are "cut out of the loop." Kurzweil emphasizes that this will have a profound, disruptive effect on human societies and on everyday life, and will mark the end of human history as we know it. In place of normal humans, Strong Artificial Intelligences and cybernetically augmented humans will become the dominant forms of sentient life on the Earth. The Singularity will be initiated once self-improving Artificial Intelligences (sentient, highly intelligent computer programs capable of quickly redesigning themselves and their offspring to improve performance) come into existence.

Underlying all of Kurzweil's ideas regarding the progress of technology and the Singularity is the Law of Accelerating Returns. The Law states that technological progress occurs exponentially instead of linearly, meaning that each new advancement enables several higher advancements instead of just one higher advancement, and concordantly, every year, more useful inventions and discoveries are made than were made in the last.

The Law of Accelerating Returns has a very important consequence in that extrapolation of exponentially improving technology trends into the future suggests, by Kurzweil's analysis, that highly advanced technologies will arrive far sooner than linear-thinking people assume. The creation of the modern Internet and the completion of the Human Genome Project are prominent examples illustrative of this point. Both were multi-year projects that relied on computer technology to reach completion. In both cases, critics derided them as hopeless since, in the beginning, both relied on computers that would have taken decades to process all of the necessary data. However, these critics had failed to take into account the exponentially improving nature of computer processing speeds and price-performance, and thus failed to see that, within a few years, the two projects would have access to vastly superior computers that would drastically shorten their timelines for completion.

Kurzweil lays forth his idea that the fate of the universe is to progress through six different epochs, characterized by the major paradigm shift that takes place after each one. The universe has already passed through four of these epochs and we are entering epoch five. Each step indirectly produces the next through chemical-, biological-, and technological Evolution (respectively).

The Six Epochs are subject to the Law of Accelerating Returns, which states that each transition occurs more rapidly than the last. Kurzweil supports this final postulate with logarithmic graphs of the chronology of important events in the history of the Universe (i.e. - the Big Bang, the origin of life, the birth of the human race, the creation of the first computer).

Epoch 1. Physics and Chemistry: This epoch starts at the beginning of the universe. In this epoch, information is mostly held in subatomic structures such as particles and atoms. That is, the most complicated stable objects in the universe do not exceed the molecular scale in size or complexity.

Epoch 2. Biology and DNA: This epoch starts with the beginning of life on Earth, suddenly giving rise to more complicated, yet stable, organisms that are capable of growth and self-sustainment. However, in this epoch organisms do not change within their lifetimes as evolution takes thousands of generations. Moreoever, in this stage, genetic information is stored in DNA molecules.

Epoch 3. Brains: The evolution of life gradually produced more and more complex organisms, necessitating the need for fast central control and thus giving rise to the evolution of brains. With brains, organisms can now change their behavior dynamically to suit changes in the environment and can also learn from past experiences. Evolutionary information is, in this stage, stored in neural patterns.

Epoch 4. Technology: Evolution of brains culminates with the evolution of humans, which possess the ability to create technology. In this stage, technological designs are also subject to evolution and information is held in hardware and software designs.

Epoch 5. The Merger of Human: Technology with Human Intelligence This epoch, which Kurzweil argues we are in the process of entering, is where technology reaches a level of sophistication and fine-structuring comparable with that of biology, allowing the two to merge to create higher forms of life and intelligence.

Epoch 6. The Universe Wakes Up: After mastering the methods of technology and biology, Kurzweil predicts that human/machine civilization will expand its frontiers into the universe, gradually (or perhaps explosively) consuming the contents of the cosmos until the universe reaches a 'saturated' state where all inanimate matter has been converted to substrates for computation and intelligence, and a truly universal super-intelligence takes form. Next, Kurzweil briefly lays out a number of postulates and consequences in the first chapter as they relate to the Singularity. As stated, the Law of Accelerating Returns implies exponential- and non-linear growth in technological capabilities and human knowledge. Information technologies are experiencing superexponential growth since human economic investment in them increases as they become more powerful.

Human brain-scanning techniques are reliant upon computers and advanced machines, which are all subject to the Law of Accelerating Returns. Therefore, our ability to understand the way the human brain works is improving tremendously every year. Based on the current trends, brain scans should give us a sound understanding of how the human brain works by the mid-2020s. Using that information, scientists should be able to create simulated human brains inside of computers, leading to the first Artificial Intelligence (a thinking computer capable of passing the Turing Test) by 2029. The computer hardware of that era should also be powerful enough to meet the cognitive needs of a simulated human mind.

Artificial Intelligence would inevitably prove superior to human intelligence since the former would combine the fast speeds, memory capacity and recall, and instant downloading/learning abilities of computers with the creativity and pattern-recognition abilities of the human mind, effectively constituting a being with the strengths of both and the weaknesses of neither. Artificial Intelligences would also have the ability to edit their own software at will to instantly improve themselves, which is an ability humans naturally lack and can only approximate through laborious genetic engineering techniques.

Biology presents inherent and insurmountable limitations that originate at the molecular level. These limitations handicap both potential mental abilities and physical abilities. This reality will inevitably lead to the creation of synthetic replacement parts that are far more durable and capable than natural human organs, limbs and cells. Some humans will choose to incorporate these engineered prostheses into their bodies, becoming cybernetically enhanced. Nanotechnology will play a major role in human brain augmentation with the advent of brain nanomachines designed specifically for interacting with human neurons. Brain cybernetics of this type would allow human users to vastly expand their cognitive abilities, to experience full-immersion virtual reality, and to directly interface with computers and other mentally augmented humans. Being naturally more powerful, the cybernetic portion of augmented human minds will come to predominate. A very broad range of advanced nanomachines will also be created, allowing the humans and machines of the future to alter their world with incredible power. Aside from the aforementioned brain augmentation role, nanotechnology could be used for a variety of tasks including a total repair of the Earth's environment and the near instantaneous manipulation of physical objects via foglets so as to blur the distinction between "real" and "virtual" reality.

Kurzweil foresees the computers of the future using three-dimensional computer chips composed of nanotubes. This architecture will allow them to operate at terahertz speeds. Computer advances like this will eventually make A.I.'s so powerful that they will completely take over the cutting edge of all scientific research and development, and will generate new advancements (including improvements to their own programming) so quickly that normal humans will not be able to follow what is going on. By this point, the Singularity will definitely have been reached. In the far future, A.I.'s and cybernetic humans will live almost exclusively in full-immersion virtual reality worlds, which will themselves be contained in advanced computers. Computers will grow ever more advanced, but at some point, the bottom limit to transistor size as defined by the laws of Physics will be reached, and performance improvements will only be possible through the creation of new computers or the expansion of existing ones. More and more inanimate matter on the Earth will be restructured to form useful computer substrate, but it will eventually prove insufficient, and the future machine race will radiate out from the Earth in all directions, "saturating" first the solar system, then the Milky Way galaxy, and eventually the entire cosmos with "intelligence" by converting unstructured, inanimate "dumb" matter (e.g., Moon rocks, dead gas giant planets, meteoroids) into structured "smart" matter that lives in the sense that it supports thinking, feeling A.I.'s. Entire celestial bodies of countless number across the universe will be totally converted into computer substrate. The length of time it will take to finish the task of "waking up" the universe depends heavily on whether the speed of light can be surpassed or circumvented—the maximum rate of travel of course limiting the speed at which the advanced Earth civilization can spread across space to new locations.

3

Important Cellular Activities

Analytical Technologies

Separation Methods

- Gas chromatography, especially when interfaced with mass spectrometry (GC-MS), is one of the most widely used and powerful methods. It offers very high chromatographic resolution, but requires chemical derivatization for many biomolecules: only volatile chemicals can be analysed without derivatization. (Some modern instruments allow '2D' chromatography, using a short polar column after the main analytical column, which increases the resolution still further.) Some large and polar metabolites cannot be analysed by GC.
- High performance liquid chromatography (HPLC). Compared to GC, HPLC has lower chromatographic resolution, but it does have the advantage that a much wider range of analytes can potentially be measured.
- Capillary electrophoresis (CE). CE has a higher theoretical separation efficiency than HPLC, and is suitable for use with a wider range of metabolite classes than is GC. As for all electrophoretic techniques, it is most appropriate for charged analytes.

Detection Methods

- Mass spectrometry (MS) is used to identify and to quantify metabolites after separation by GC, HPLC (LC-MS), or CE. GC-MS is the most 'natural' combination of the three, and was the first to be developed. In addition, mass spectral fingerprint libraries exist or can be developed that allow identification of a metabolite according to its fragmentation pattern. MS is both sensitive (although, particularly for HPLC-MS, sensitivity is more of an issue as it is affected by the charge on the metabolite, and can be subject to ion suppression artifacts) and can be very specific. There are also a number of studies which use MS as a stand-alone technology: the sample is infused directly into the mass spectrometer with no prior separation, and the MS serves to both separate and to detect metabolites.

- Surface-based mass analysis has seen a resurgence in the past decade, with new MS technologies focused on increasing sensitivity, minimizing background, and reducing sample preparation. The ability to analyze metabolites directly from biofluids and tissues continues to challenge current MS technology, largely because of the limits imposed by the complexity of these samples, which contain thousands to tens of thousands of metabolites. Among the technologies being developed to address this challenge is Nanostructure-Initiator MS (NIMS), a desorption/ionization approach that does not require the application of matrix and thereby facilitates small-molecule (i.e., metabolite) identification.

  MALDI is also used however, the application of a MALDI matrix can add significant background at <1000 Da that complicates analysis of the low-mass range (i.e., metabolites). In addition, the size of the resulting matrix crystals limits the spatial resolution that can be achieved in tissue imaging. Because of these limitations, several other matrix-free desorption/ionization approaches have been applied to the analysis of biofluids and tissues. Secondary ion mass spectrometry (SIMS) was one of the first matrix-free desorption/ionization approaches used to analyze metabolites from biological samples. SIMS uses a high-energy primary ion beam to desorb and generate secondary ions from a surface. The primary advantage of SIMS is its high spatial resolution (as small as 50 nm), a powerful characteristic for tissue imaging with MS. However, SIMS has yet to be readily applied to the analysis of biofluids and tissues because of its limited sensitivity at >500 Da and analyte fragmentation generated by the high-energy primary ion beam. Desorption electrospray ionization (DESI) is a matrix-free technique for analyzing biological samples that uses a charged solvent spray to desorb ions from a surface. Advantages of DESI are that no special surface is required and the analysis is performed at ambient pressure with full access to the sample during acquisition. A limitation of DESI is spatial resolution because "focusing" the charged solvent spray is difficult. However, a recent development termed laser ablation ESI (LAESI) is a promising approach to circumvent this limitation.

- Nuclear magnetic resonance (NMR) spectroscopy. NMR is the only detection technique which does not rely on separation of the analytes, and the sample can thus be recovered for further analyses. All kinds of small molecule metabolites can be measured simultaneously - in this sense, NMR is close to being a universal detector. The main advantages of NMR are high analytical reproducibility and simplicity of sample preparation. Practically, however, it is relatively insensitive compared to mass spectrometry-based techniques.
- Although NMR and MS are the most widely used techniques, other methods of detection that have been used include ion-mobility spectrometry, electrochemical detection (coupled to HPLC) and radiolabel (when combined with thin-layer chromatography).

Statistical Methods

The data generated in metabolomics usually consist of measurements performed on subjects under various conditions. These measurements may be digitized spectra, or a list of metabolite levels. In its simplest form this generates a matrix with rows corresponding to subjects and columns corresponding to metabolite levels. Several statistical programs are currently available for analysis of both NMR and mass spectrometry data. For mass spectrometry data, software is available that identifies molecules that vary in subject groups on the basis of mass and sometimes retention time depending on the experimental design. The first comprehensive software to analyze global mass spectrometry-based metabolomics datasets was developed by the Siuzdak laboratory at The Scripps Research Institute in 2006.

This software, called XCMS, is freely available, has over 20,000 downloads since its inception in 2006, and is one of the most widely cited mass spectrometry-based metabolomics software programs in scientific literature. Other popular metabolomics programs for mass spectral analysis are MZmine, MetAlign, MathDAMP, which also compensate for retention time deviation during sample analysis. LCMStats is another R package for detailed analysis of liquid chromatography mass spectrometry (LCMS) data and is helpful in identification of co-eluting ions especially isotopologues from a complicated metabolic profile. It combines xcms package functions and can be used to apply many statistical functions for correcting detector saturation using coates correction and creating heat plots. Metabolomics data may also be analyzed by statistical projection (chemometrics) methods such as principal components analysis and partial least squares regression.

Key Applications

- Toxicity assessment/toxicology. Metabolic profiling (especially of urine or blood plasma samples) can be used to detect the physiological changes caused by toxic insult of a chemical (or mixture of chemicals). In many cases, the observed changes can be related to specific syndromes, e.g. a specific lesion in liver or kidney. This is of particular relevance to pharmaceutical companies wanting to test the toxicity of potential drug candidates: if a compound can be eliminated before it reaches clinical trials on the grounds of adverse toxicity, it saves the enormous expense of the trials.
- Functional genomics. Metabolomics can be an excellent tool for determining the phenotype caused by a genetic manipulation, such as gene deletion or insertion. Sometimes this can be a sufficient goal in itself—for instance, to detect any phenotypic changes in a genetically-modified plant intended for human or animal consumption. More exciting is the prospect of predicting the function of unknown genes by comparison with the metabolic perturbations caused by deletion/insertion of known genes. Such advances are most likely to come from model organisms such as Saccharomyces cerevisiae and Arabidopsis thaliana. The Cravatt laboratory at The Scripps Research Institute has recently applied this technology to mammalian systems, identifying the N-acyltaurines as previously uncharacterized endogenous substrates for the enzyme fatty acid amide hydrolase (FAAH) and the monoalkylglycerol ethers (MAGEs) as endogenous substrates for the uncharacterized hydrolase KIAA1363.
- Nutrigenomics is a generalised term which links genomics, transcriptomics, proteomics and metabolomics to human nutrition. In general a metabolome in a given body fluid is influenced by endogenous factors such as age, sex, body composition and genetics as well as underlying pathologies. The large bowel microflora are also a very significant potential confounder of metabolic profiles and could be classified as either an endogenous

or exogenous factor. The main exogenous factors are diet and drugs. Diet can then be broken down to nutrients and non-nutrients. Metabolomics is one means to determine a biological endpoint, or metabolic fingerprint, which reflects the balance of all these forces on an individual's metabolism.

Environmental Metabolomics

- Environmental Metabolomics is the application of metabolomics to characterise the interactions of organisms with their environment. This approach has many advantages for studying organism–environment interactions and for assessing organism function and health at the molecular level. As such, metabolomics is finding an increasing number of applications in the environmental sciences, ranging from understanding organismal responses to abiotic pressures, to investigating the responses of organisms to other biota. These interactions can be studied from individuals to populations, which can be related to the traditional fields of ecophysiology and ecology, and from instantaneous effects to those over evolutionary time scales, the latter enabling studies of genetic adaptation.

Lipidomics

Lipidomics may be defined as the large-scale study of pathways and networks of cellular lipids in biological systems The word "lipidome" is used to describe the complete lipid profile within a cell, tissue or organism and is a subset of the "metabolome" which also includes the three other major classes of biological molecules: proteins/amino-acids, sugars and nucleic acids.

Lipidomics is a relatively recent research field that has been driven by rapid advances in technologies such as mass spectrometry (MS), nuclear magnetic resonance (NMR) spectroscopy, fluorescence spectroscopy, dual polarisation interferometry and computational methods, coupled with the recognition of the role of lipids in many metabolic diseases such as obesity, atherosclerosis, stroke, hypertension and diabetes.

This rapidly expanding field complements the huge progress made in genomics and proteomics, all of which constitute the family of systems biology.

Lipidomics research involves the identification and quantification of the thousands of cellular lipid molecular species and their interactions with other lipids, proteins, and other metabolites. Investigators in lipidomics examine the structures, functions, interactions, and dynamics of cellular lipids and the changes that occur during perturbation of the system. Han and Gross first defined the field of lipidomics through integrating the specific chemical properties inherent in lipid molecular species with a comprehensive mass spectrometric approach.

Although lipidomics is under the umbrella of the more general field of "metabolomics", lipidomics is itself a distinct discipline due to the uniqueness and functional specificity of lipids relative to other metabolites.

In lipidomic research, a vast amount of information quantitatively describing the spatial and temporal alterations in the content and composition of different lipid molecular species is accrued after perturbation of a cell through changes in its physiological or pathological state.

Information obtained from these studies facilitates mechanistic insights into changes in cellular function. Therefore, lipidomic studies play an essential role in defining the biochemical mechanisms of lipid-related disease processes through identifying alterations in cellular lipid metabolism, trafficking and homeostasis. The growing attention on lipid research is also seen from the initiatives underway of the LIPID Metabolites And Pathways Strategy (LIPID MAPS Consortium). and The European Lipidomics Initiative (ELIfe).

Structural Diversity of Lipids

Lipids are a diverse and ubiquitous group of compounds which have many key biological functions, such as acting as structural components of cell membranes, serving as energy storage sources and participating in signaling pathways.

Lipids may be broadly defined as hydrophobic or amphipathic small molecules that originate entirely or in part from two distinct types of biochemical subunits or "building blocks": ketoacyl and isoprene groups.

The huge structural diversity found in lipids arises from the biosynthesis of various combinations of these building blocks. For example, glycerophospholipids are composed of a glycerol backbone linked to one of approximately 10 possible headgroups and also to 2 fatty acyl/alkyl chains, which in turn may have 30 or more different molecular structures. In practice, not all possible permutations are detected experimentally, due to chain preferences depending on the cell type and also to detection limits - nevertheless several hundred distinct glycerophospholipid molecular species have been detected in mammalian cells.

Experimental Techniques

Lipid Extraction

Most methods of lipid extraction and isolation from biological samples exploit the high solubility of hydrocarbon chains in organic solvents. Given the diversity in lipid classes, it is not possible to accommodate all classes with a common extraction method. The traditional Bligh/ Dyer procedure uses chloroform/methanol-based protocols that include phase partitioning into the organic layer.

These protocols work relatively well for a wide variety of physiologically relevant lipids but they have to be adapted for complex lipid chemistries and low-abundance and labile lipid metabolites When organic soil was used ,citrate buffer in the extraction mixture gave higher amounts of lipid phosphate than acetate buffer,Tris,H2O or phosphate buffer.

Lipid Separation

The simplest method of lipid separation is the use of thin layer chromatography (TLC). Although not as sensitive as other methods of lipid detection, it offers a rapid and comprehensive screening tool prior to more sensitive and sophisticated techniques. Solid-phase extraction (SPE) chromatography is useful for rapid, preparative separation of crude lipid mixtures into different lipid classes. This involves the use of prepacked columns containing silica or other stationary phases to separate glycerophospholipids, fatty acids, cholesteryl esters, glycerolipids, and sterols from crude lipid mixtures. High performance liquid chromatography (HPLC or LC) is extensively used in lipidomic analysis to separate lipids prior to mass analysis. Separation can be achieved by either normal-phase HPLC or reverse-phase HPLC.

For example, normal phase HPLC effectively separates glycerophospholipids on the basis of headgroup polarity, whereas reverse-phase HPLC effectively separates fatty acids such as eicosanoids on the basis of chain length, degree of unsaturation and substitution. HPLC of lipids may either be performed offline or online where the eluate is integrated with the ionization source of a mass spectrometer.

Lipid Detection

The progress of modern lipidomics has been greatly accelerated by the development of spectrometric methods in general and soft ionization techniques for mass spectrometry such as electrospray ionization (ESI) and matrix-assisted laser desorption/ionization (MALDI) in particular. "Soft" ionization does not cause extensive fragmentation, so that comprehensive detection of an entire range of lipids within a complex mixture can be correlated to experimental conditions or disease state. In addition to ESI and MALDI, the technique of atmospheric pressure chemical ionization (APCI) has become increasingly popular for the analysis of nonpolar lipids.

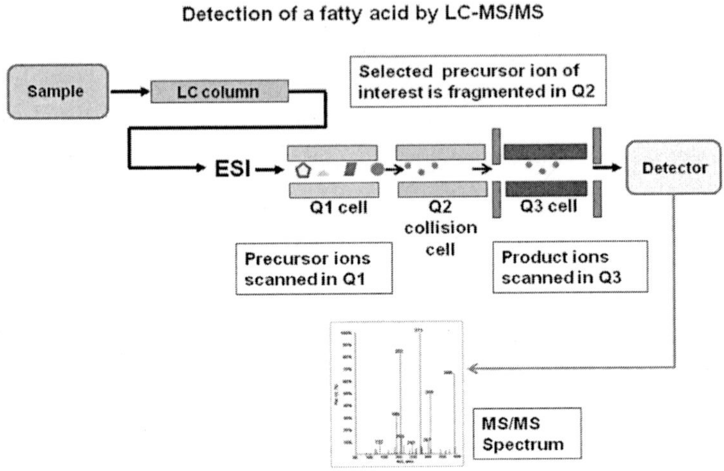

Figure 1: *Schema showing detection of a fatty acid by LC-MS/MS using a linear ion-trap instrument and an electrospray (ESI) ion source.*

ESI MS

ESI-MS was initially developed by Fenn and colleagues for analysis of biomolecules. It depends on the formation of gaseous ions from polar, thermally labile and mostly non-volatile molecules and thus is completely suitable for a variety of lipids. It is a soft-ionization method that rarely disrupts the chemical nature of the analyte prior to mass analysis. Various ESI-MS methods have been developed for analysis of different classes, subclasses, and individual lipid species from biological extracts. Comprehensive reviews of the methods and their application have recently been published. The major advantages of ESI-MS are high accuracy, sensitivity, reproducibility, and the applicability of the technique to complex solutions without prior derivatization. Han and coworkers have developed a method known as"shotgun lipidomics" which involves direct infusion of a crude lipid extract into an ESI source optimized for intrasource separation of lipids based on their intrinsic electrical properties.

MALDI MS

MALDI mass spectrometry is a laser-based soft-ionization method often used for analysis of large proteins, but has been used successfully for lipids. The lipid is mixed with a matrix, such as 2,5-dihydroxybenzoic acid, and applied to a sample holder as a small spot. A laser is fired at the spot, and the matrix absorbs the energy, which is then transferred to the analyte, resulting in ionization of the molecule. MALDI-Time-of-flight (MALDI-TOF) MS has become a very promising approach for lipidomics studies, particularly for the imaging of lipids from tissue slides.

APCI MS

The source for APCI is similar to ESI except that ions are formed by the interaction of the heated analyte solvent with a corona discharge needle set at a high electrical potential. Primary ions are formed immediately surrounding the needle, and these interact with the solvent to form secondary ions that ultimately ionize the sample. APCI is particularly useful for the analysis of nonpolar lipids such as triacylglycerols, sterols, and fatty acid esters.

Imaging Techniques

Recent developments in MALDI methods have enabled direct detection of lipids in-situ. Abundant lipid-related ions are produced from the direct analysis of thin tissue slices when sequential spectra are acquired across a tissue surface that has been coated with a MALDI matrix. Collisional activation of the molecular ions can be used to determine the lipid family and often structurally define the molecular species. This technique enables detection of phospholipids, sphingolipids and glycerolipids in tissues such as heart, kidney and brain.

Lipidomic Profiling

Lipid profiling is a targeted metabolomics platform that provides a comprehensive analysis of lipid species within a cell or tissue. Profiling based on electrospray ionization tandem mass spectrometry (ESI-MS/MS) is capable of providing quantitative data and is adaptable to high throughput analyses.

The powerful approach of transgenics, namely deletion and/or overexpression of a gene product coupled with lipidomics, can give valuable insights into the role of biochemical pathways.

Lipid profiling techniques have also been applied to plants and microorganisms such as yeast. A combination of quantitative lipidomic data in conjunction with the corresponding transcriptional data (using gene-array methods) and proteomic data (using tandem MS) enables a systems biology approach to a more in-depth understanding of the metabolic or signaling pathways of interest.

Informatics

A major challenge for lipidomics, in particular for MS-based approaches, lies in the computational and bioinformatic demands of handling the large amount of data that arise at various stages along the chain of information acquisition and processing. Chromatographic and MS data collection requires substantial efforts in spectral alignment and statistical evaluation of fluctuations in signal intensities. Such variations have a multitude of origins, including biological variations, sample handling and analytical accuracy. As a consequence several replicates are normally required for reliable determination of lipid levels in complex mixtures. Within the last few years, a number of software packages have been developed by various companies and research groups to analyze data generated by MS profiling of metabolites, including lipids.

The data processing for differential profiling usually proceed through several stages, including input file manipulation, spectral filtering, peak detection, chromatographic alignment, normalization, visualization, and data export. An example of metabolic profiling software is the freely-available Java-based Mzmine application.

Some software packages such as Markerview include multivariate statistical analysis (for example, principal component analysis) and these will be helpful for the identification of correlations in lipid metabolites that are associated with a physiological phenotype, in particular for the development of lipid-based biomarkers.Another objective of the information technology side of lipidomics involves the construction of metabolic maps from data on lipid structures and lipid-related protein and genes.

Some of these lipid pathways are extremely complex, for example the mammalian glycosphingolipid pathway. The establishment of searchable and interactive databases of lipids and lipid-related genes/proteins is also an extremely important resource as a reference for the lipidomics community. Integration of these databases with MS and other experimental data, as well as with metabolic networks offers an opportunity to devise therapeutic strategies to prevent or reverse these pathological states involving dysfunction of lipid-related processes.

Shotgun Proteomics

Shotgun proteomics is a method of identifying proteins in complex mixtures using a combination of high performance liquid chromatography combined with mass spectrometry. The name is derived from shotgun sequencing of DNA which is itself named by analogy with the rapidly-expanding, quasi-random firing pattern of a shotgun. In shotgun proteomics, the proteins in the mixture are digested and the resulting peptides are separated by liquid chromatography. Tandem mass spectrometry is then used to identify the peptides.

Top-down Proteomics

Top-down proteomics is a method of protein identification that uses an ion trapping mass spectrometer to store an isolated protein ion for mass measurement and tandem mass spectrometry analysis. The name is derived from the similar approach to DNA seqencing. Proteins are typically ionized by electrospray ionization and trapped in a Fourier transform ion cyclotron resonance (Penning trap) or quadrupole ion trap (Paul trap) mass spectrometer. Fragmentation for tandem mass spectrometry is accomplished by electron-capture dissociation or electron-transfer dissociation.

Recently, a top-down approach has been developed for mapping the connectivity of the disulfide-rich peptide hedyotide B2 . This method allows rapid characterization of the disulfide pattern of cystine-knot miniproteins such as cyclotides, conotoxin, knottin and plant defensins.

Bottom-up Proteomics

Bottom-up proteomics is a common method to identify proteins and characterize their amino acid sequences and post-translational modifications by proteolytic digestion of proteins prior to analysis by mass spectrometry. The proteins may first be purified by a method such as gel electrophoresis resulting in one or a few proteins in each proteolytic digest. Alternatively, the crude protein extract is digested directly, followed by one or more dimensions of separation of the peptides by liquid chromatography coupled to mass spectrometry, a technique known as shotgun proteomics. By comparing the masses of the proteolytic peptides or their tandem mass spectra with those predicted from a sequence database or annotated peptide spectral in a peptide spectral library, peptides can be identified and multiple peptide identifications assembled into a protein identification.

Transcriptomics

The transcriptome is the set of all RNA molecules, including mRNA, rRNA, tRNA, and other non-coding RNA produced in one or a population of cells.

Scope

The term can be applied to the total set of transcripts in a given organism, or to the specific subset of transcripts present in a particular cell type. Unlike the genome, which is roughly fixed for a given cell line (excluding mutations), the transcriptome can vary with external environmental conditions. Because it includes all mRNA transcripts in the cell, the transcriptome reflects the genes that are being actively expressed at any given time, with the exception of mRNA degradation phenomena such as transcriptional attenuation. The study of transcriptomics, also referred to as expression profiling, examines the expression level of mRNAs in a given cell population, often using high-throughput techniques based on DNA microarray technology. The use of next-generation sequencing technology to study the transcriptome at the nucleotide level is known as RNA-Seq.

Analysis

DNA microarrays can provide a method for comparing on a genome-wide basis the abundance of DNAs in the same samples. The DNA in spots can only be PCR products that are specific for individual genes.

A DNA copy of RNA is made using the enzyme reverse transcriptase. Gene arrays can only be used to investigate genes in eukaryotic cells. Sequencing is now being used instead of gene arrays to quantify DNA levels, at least semi quantitatively.

A number of organism-specific transcriptome databases have been constructed and annotated to aid in the identification of genes that are differentially expressed in distinct cell populations or subtypes.

Applications

The transcriptomes of stem cells and cancer cells are of particular interest to researchers who seek to understand the processes of cellular differentiation and carcinogenesis.

Analysis of the transcriptomes of human oocytes and embryos is used to understand the molecular mechanisms and signaling pathways controlling early embryonic development, and could theoretically be a powerful tool in making proper embryo selection in in vitro fertilisation.

Relation to Proteome

Proteome

The transcriptome can be seen as a precursor for the proteome, that is, the entire set of proteins expressed by a genome. However, the analysis of relative mRNA expression levels can be complicated by the fact that relatively small changes in mRNA expression can produce large changes in the total amount of the corresponding protein present in the cell. One analysis method, known as Gene Set Enrichment Analysis, identifies coregulated gene networks rather than individual genes that are up- or down-regulated in different cell populations. Although microarray studies can reveal the relative amounts of different mRNAs in the cell, levels of mRNA are not directly proportional to the expression level of the proteins they code for.

The number of protein molecules synthesized using a given mRNA molecule as a template is highly dependent on translation-initiation features of the mRNA sequence; in particular, the ability of the translation initiation sequence is a key determinant in the recruiting of ribosomes for protein translation. The complete protein complement of a cell or organism is known as the proteome. A study of 158,807 mouse transcripts revealed that 4,520 of these transcripts form antisense partners that are base pair complementary to the exons of genes. These results raise the possibility that significant numbers of "antisense RNA-coding genes" might participate in the regulation of the levels of expression of protein-coding mRNAs.

Phosphoproteomics

Phosphoproteomics is a branch of proteomics that identifies, catalogs, and characterizes proteins containing a phosphate group as a post-translational modification. Phosphorylation is a key reversible modification that regulates protein function, subcellular localization, complex formation, degradation of proteins and therefore cell signalling networks. With all of these modification results, it is assumed that up to 30% of all proteins may be phosphorylated, some multiple times.

Compared to expression analysis, phosphoproteomics provides two additional layers of information. First, it provides clues on what protein or pathway might be activated because a change in phosphorylation status almost always reflects a change in protein activity. Second, it indicates what proteins might be potential drug targets as exemplified by the kinase inhibitor Gleevec. While phosphoproteomics will greatly expand knowledge about the numbers and types of phosphoproteins, its greatest promise is the rapid analysis of entire phosphorylation based signalling networks.

Overview of Phosphoproteomic Analysis

A sample large-scale phosphoproteomic analysis includes

1. Cultured cells undergo SILAC encoding.
2. Cells are stimulated with factor of interest (eg. growth factor, hormone).
3. Stimulation can occur for various lengths of time for temporal analysis.

4. Cells are lysed and enzymatically digested.
5. Peptides are separated using ion exchange chromatography.
6. Phosphopeptides are enriched using phosphospecific antibodies, immobilized metal affinity chromatography or titanium dioxide (TiO_2) chromatography.
7. Phosphopeptides are analyzed using mass spectrometry.
8. Peptides are sequenced and analyzed.

Tools and Methods

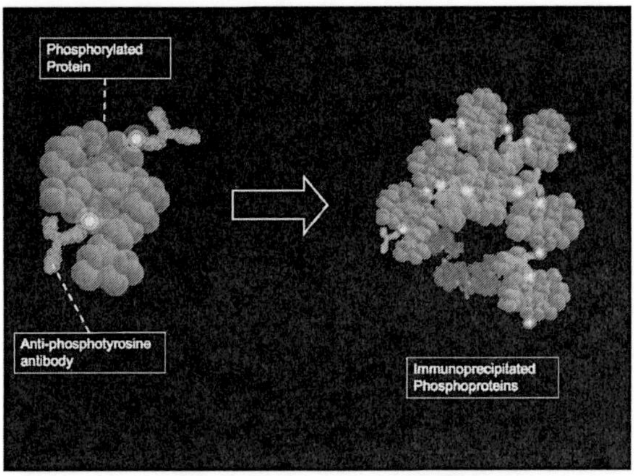

Figure 2: *Method of phosphoprotein purification by immunoprecipitation with anti-phosohotyrosine antibodies.*

The analysis of the entire complement of phosphorylated proteins in a cell is certainly a feasible option. This is due to the optimization of enrichment protocols for phosphoproteins and phosphopeptides, better fractionation techniques using chromatography, and improvement of methods to selectively visualize phosphorylated residues using mass spectrometry. Although the current procedures for phosphoproteomic analysis are greatly improved, there is still sample loss and inconsistencies with regards to sample preparation, enrichment , and instrumentation. Bioinformatics tools and biological sequence databases are also necessary for high-throughput phosphoproteomic studies.

Enrichment Strategies

Previous procedures to isolate phosphorylated proteins included radioactive labeling with P-labeled ATP followed by SDS polyacrylamide gel electrophoresis or thin layer chromatography. These traditional methods are inefficient because it is impossible to obtain large amounts of proteins required for phosphorylation analysis. Therefore, the current and simplest methods to enrich phosphoproteins are affinity purification using phosphospecific antibodies, immobilized metal affinity chromatography (IMAC), strong cation exchange (SCX) chromatography, or titanium dioxide chromatography. Antiphosphotyrosine antibodies have been proven very successful in purification, but fewer reports have been published using antibodies against phosphoserine- or phosphothreonine-containing proteins. IMAC enrichment is based on phosphate affinity for immobilized metal chelated to the resin. SCX separates phosphorylated from non-phosphorylated peptides based on the negatively charged phosphate group. Titanium dioxide chromatography is a newer technique that requires significantly less column preparation time. Many phosphoproteomic studies use a combination of these enrichment strategies to obtain the purest sample possible.

Mass Spectrometry Analysis

Mass spectrometry is currently the best method to adequately compare pairs of protein samples. The two main procedures to perform this task are using isotope-coded affinity tags (ICAT) and stable isotopic amino acids in cell culture (SILAC). In the ICAT procedure samples are labeled individually after isolation with mass-coded reagents that modify cysteine residues. In SILAC, cells are cultured separately in the presence of different isotopically labeled amino acids for several cell divisions allowing cellular proteins to incorporate the label. Mass spectrometry is subsequently used to identify phosphoserine, phosphothreonine, and phosphotyrosine-containing peptides.

Phosphoproteomics in the Study of Signal Transduction

Intracellular signal transduction is primarily mediated by the reversible phosphorylation of various signalling molecules by enzymes dubbed kinases. Kinases transfer phosphate groups from ATP to specific serine, threonine or tyrosine residues of target molecules. The resultant phosphorylated protein may have altered activity level, subcellular localization or tertiary structure.

Phosphoproteomic analyses are ideal for the study of the dynamics of signalling networks. In one study design, cells are exposed to SILAC labelling and then stimulated by a specific growth factor. The cells are collected at various timepoints, and the lysates are combined for analysis by tandem MS. This allows experimenters to track the phosphorylation state of many phosphoproteins in the cell over time. The ability to measure the global phosphorylation state of many proteins at various time points makes this approach much more powerful than traditional biochemical methods for analyzing signalling network behavior. One study was able to simultaneously measure the fold-change in phosphorylation state of 127 proteins between unstimulated and EphrinB1-stimulated cells. Of these 127 proteins, 40 showed increased phosphorylation with stimulation by EphrinB1. The researchers were able to use this information in combination with previously published data to construct a signal transduction network for the proteins downstream of the EphB2 receptor.

Another recent phosphoproteomic study included large-scale identification and quantification of phosphorylation events triggered by the anti-diuretic hormone vasopressin in kidney collecting duct. A total of 714 phosphorylation sites on 223 unique phosphoproteins were identified, including three novel phosphorylation sites in the vasopressin-sensitive water channel aquaporin-2 (AQP2).

Phosphoproteomics in the Study of Cancer

Since the inception of phosphoproteomics, cancer research has focused on changes to the phosphoproteome during tumor development. Phosphoproteins could be cancer markers useful to cancer diagnostics and therapeutics. In fact, research has shown that there are distinct phosphotyrosine proteomes of breast and liver tumors. There is also evidence of hyperphosphorylation at tyrosine residues in breast tumors but not in normal tissues. Findings like these suggest that it is possible to mine the tumor phosphoproteome for potential biomarkers.

Increasing amounts of data are available suggesting that distinctive phosphoproteins exist in various tumors and that phosphorylation profiling could be used to fingerprint cancers from different origins. In addition, systematic cataloguing of tumor-specific phosphoproteins in individual patients could reveal multiple causative players during cancer formation. By correlating this experimental data to clinical data such as drug response and disease outcome, potential cancer markers could be identified for diagnosis, prognosis, prediction of drug response, and potential drug targets.

Limitations

While phosphoproteomics has greatly expanded knowledge about the numbers and types of phosphoproteins, along with their role in signaling networks, there are still a few limitations to these techniques. To begin with, isolation methods such as anti-phosphotyrosine antibodies do not distinguish between isolating tyrosine-phosphorylated proteins and proteins associated with tyrosine-phosphorylated proteins. Therefore, even though phosphorylation dependent protein-protein interactions are very important, it is important to remember that a protein detected by this method is not necessarily a direct substrate of any tyrosine kinase. Only by digesting the samples before immunoprecipitation can isolation of only phosphoproteins and temporal profiles of individual phosphorylaiton sites be produced.

Another limitation is that some relevant proteins will likely be missed since no extraction condition is all encompassing. It is possible that proteins with low stoichiometry of phosphorylation, in very low abundance, or phosphorylated as a target for rapid degradation will be lost.

PEGylation

PEGylation is the process of covalent attachment of polyethylene glycol polymer chains to another molecule, normally a drug or therapeutic protein. PEGylation is routinely achieved by incubation of a reactive derivative of PEG with the target macromolecule. The covalent attachment of PEG to a drug or therapeutic protein can "mask" the agent from the host's immune system (reduced immunogenicity and antigenicity), and increase the hydrodynamic size (size in solution) of the agent which prolongs its circulatory time by reducing renal clearance. PEGylation can also provide water solubility to hydrophobic drugs and proteins.

History

In 1970s, pioneering research by Dr. Frank Davis, Dr. Abraham Abuchowski and colleagues foresaw the potential of the conjugation of Polyethylene glycol (PEG) to proteins. Dr. Abuchowski founded Enzon, Inc, which brought three PEGylated drugs to market. He is the founder and president of Prolong Pharmaceuticals.

PEGylation, is a process of attaching the strands of the polymer PEG to molecules most typically peptides, proteins, and antibody fragments, that can help to meet the challenges of improving the safety and efficiency of many therapeutics.

It produces alterations in the physiochemical properties including changes in conformation, electrostatic binding, hydrophobicity etc. These physical and chemical changes increase systemic retention of the therapeutic agent. Also, it can influence the binding affinity of the therapeutic moiety to the cell receptors and can alter the absorption and distribution patterns.

PEGylation, by increasing the molecular weight of a molecule, can impart several significant pharmacological advantages over the unmodified form, such as:

- Improved drug solubility
- Reduced dosage frequency, without diminished efficacy with potentially reduced toxicity
- Extended circulating life
- Increased drug stability
- Enhanced protection from proteolytic degradation

PEGylated drugs have the following commercial advantages also:

- Opportunities for new delivery formats and dosing regimens
- Extended patent life of previously approved drugs

PEGylated Pharmaceuticals on the Market

The clinical value of PEGylation is now well established. ADAGEN (PEG- bovine adenosine deaminase) manufactured by Enzon Pharmaceuticals, Inc., US was the first PEGylated protein approved by the U.S. Food and Drug Administration (FDA) in March 1990, to enter the market.

It is used to treat X-linked severe combined immunogenicity syndrome, as an alternative to bone marrow transplantation and enzyme replacement by gene therapy.

Since the introduction of ADAGEN, a large number of PEGylated protein and peptide pharmaceuticals have followed and many others are under clinical trial or under development stages. Some of the successful examples are:

- PEGASYS: PEGylated interferon alpha for use in the treatment of chronic hepatitis C and hepatitis B (Hoffmann-La Roche)
- Pegintron: PEGylated interferon alpha for use in the treatment of chronic hepatitis C and hepatitis B (Schering-Plough / Enzon)
- Oncaspar: PEGylated L-asparaginase for the treatment of acute lymphoblastic leukemia in patients who are hypersensitive to the native unmodified form of L-asparaginase (Enzon). This drug was recently approved for front line use.
- Neulasta: PEGylated recombinant methionyl human granulocyte colony-stimulating factor for severe cancer chemotherapy-induced neutropenia (Amgen)
- Doxil/Caelyx: PEGylated liposome containing doxorubicin for the treatment of cancer (Ortho Biotech / Schering-Plough)
- Pegloticase (Krystexxa) - PEGylated Uricase for the treatment of gout (Savient)

PEG Moiety Properties

PEG is a particularly attractive polymer for conjugation. The specific characteristics of PEG moieties relevant to pharmaceutical applications are:

- Water solubility
- High mobility in solution
- Lack of toxicity and immunogenicity
- Ready clearance from the body
- Altered distribution in the body

PEGylation Process

The first step of the PEGylation is the suitable functionalization of the PEG polymer at one or both terminals. PEGs that are activated at each terminus with the same reactive moiety are known as "homobifunctional", whereas if the functional groups present are different, then the PEG derivative is referred as "heterobifunctional" or "heterofunctional." The chemically active or activated derivatives of the PEG polymer are prepared to attach the PEG to the desired molecule.

The overall PEGylation processes used to date for protein conjugation can be broadly classified into two types, namely a solution phase batch process and an on-column fed-batch process. The simple and commonly adopted batch process involves the mixing of reagents together in a suitable buffer solution, preferably at a temperature between 4 and 6°C, followed by the separation and purification of the desired product using a suitable technique based on its physicochemical properties, including size exclusion chromatography (SEC), ion exchange chromatography (IEX), hydrophobic interaction chromatography (HIC) and membranes or aqueous two phase systems.

The choice of the suitable functional group for the PEG derivative is based on the type of available reactive group on the molecule that will be coupled to the PEG. For proteins, typical reactive amino acids include lysine, cysteine, histidine, arginine, aspartic acid, glutamic acid, serine, threonine, tyrosine. The N-terminal amino group and the C-terminal carboxylic acid can also be used as a site specific site by conjugation with aldehyde functional polymers.

The techniques used to form first generation PEG derivatives are generally reacting the PEG polymer with a group that is reactive with hydroxyl groups, typically anhydrides, acid chlorides, chloroformates and carbonates. In the second generation PEGylation chemistry more efficient functional groups such as aldehyde, esters, amides etc made available for conjugation. As applications of PEGylation have become more and more advanced and sophisticated, there has been an increase in need for heterobifunctional PEGs for conjugation. These heterobifunctional PEGs are very useful in linking two entities, where a hydrophilic, flexible and biocompatible spacer is needed. Preferred end groups for heterobifunctional PEGs are maleimide, vinyl sulfones, pyridyl disulfide, amine, carboxylic acids and NHS esters.

Third generation pegylation agents, where the shape of the polymer has been branched, Y shaped or comb shaped are available which show reduced viscosity and lack of organ accumulation.

Future Perspectives

Even though nearly four decades of development in PEGylation technology has proven its pharmacological advantages and acceptability, the technology still lags in providing a commercially attractive, generic process to produce highly specific PEGylated therapeutic products at high yield. As a multi-million dollar annual business with growing interest from both emerging biotechnology and established multinational pharmaceutical companies, there is great scientific and commercial interest in improving present methodologies and in introducing innovative process variations.

Activity Based Proteomics

Activity based proteomics, or activity based protein profiling (ABPP) is a functional proteomic technology that uses specially designed chemical probes that react with mechanistically-related classes of enzymes. The basic unit of ABPP is the probe which typically consists of two elements: a reactive group (RG) and a tag. Additionally, some probes may contain a binding group which enhances selectivity. The reactive group usually contains an electrophile that gets covalently-linked to a nucleophilic residue in the active site of an active enzyme. An enzyme that is inhibited by enzyme inhibitors or post-translational modifications will not react with an activity-based probe. The tag may be either a reporter such as a fluorophore or an affinity label such as biotin or an alkyne or azide for use with the Huisgen 1,3-dipolar cycloaddition (also known as click chemistry) .

A major advantage of ABPP is the ability to monitor the availability of the enzyme active site directly, rather than being limited to protein or mRNA abundance. With classes of enzymes such as the serine proteases and metalloproteases that often interact with endogenous inhibitors or that exist as inactive zymogens, this technique offers a valuable advantage over traditional techniques that rely on abundance rather than activity. Finally, in recent years ABPP has been combined with tandem mass spectrometry enabling the identification of hundreds of active enzymes from a single sample. This technique, known as ABPP-MudPIT is especially useful for profiling inhibitor selectivity as the potency of an inhibitor can be tested against hundreds of targets simultaneously. ABPP was originally invented by James C. Powers (Georgia Tech, Atlanta) in the early 1990s and subsequently adopted and developed by the other labs, for example Benjamin Cravatt III at The Scripps Research Institute, Matthew Bogyo at Stanford, and others.

Protein Databases

UniProt

UniProt is a comprehensive, high-quality and freely accessible database of protein sequence and functional information, many of which are derived from genome sequencing projects. It contains a large amount of information about the biological function of proteins derived from the research literature.

The UniProt Consortium

The UniProt Consortium comprises the European Bioinformatics Institute (EBI), the Swiss Institute of Bioinformatics (SIB), and the Protein Information Resource (PIR). EBI, located at the Wellcome Trust Genome Campus in Hinxton, UK, hosts a large resource of bioinformatics databases and services.

SIB, located in Geneva, Switzerland, maintains the ExPASy (Expert Protein Analysis System) servers that are a central resource for proteomics tools and databases.

PIR, hosted by the National Biomedical Research Foundation (NBRF) at the Georgetown University Medical Center in Washington, DC, USA, is heir to the oldest protein sequence database, Margaret Dayhoff's Atlas of Protein Sequence and Structure, first published in 1965. In 2002, EBI, SIB, and PIR joined forces as the UniProt Consortium.

The Roots of UniProt Databases

Each consortium member is heavily involved in protein database maintenance and annotation. Until recently, EBI and SIB together produced the Swiss-Prot and TrEMBL databases, while PIR produced the Protein Sequence Database (PIR-PSD). These databases coexisted with differing protein sequence coverage and annotation priorities. Swiss-Prot was created in 1986 by Amos Bairoch during his PhD and developed by the Swiss Institute of Bioinformatics and subsequently developed by Rolf Apweiler at the European Bioinformatics Institute. Swiss-Prot aimed to provide reliable protein sequences associated with a high level of annotation (such as the description of the function of a protein, its domain structure, post-translational modifications, variants, etc.), a minimal level of redundancy and high level of integration with other databases.

Recognizing that sequence data were being generated at a pace exceeding Swiss-Prot's ability to keep up, TrEMBL (Translated EMBL Nucleotide Sequence Data Library) was created to provide automated annotations for those proteins not in Swiss-Prot. Meanwhile, PIR maintained the PIR-PSD and related databases, including iProClass, a database of protein sequences and curated families.

The consortium members pooled their overlapping resources and expertise, and launched UniProt in December 2003.

Organization of UniProt Databases

UniProt provides four core databases:

UniProtKB

UniProt Knowledgebase (UniProtKB) is a protein database curated by experts, consisting of two sections. UniProtKB/Swiss-Prot (containing reviewed, manually annotated entries) and UniProtKB/TrEMBL (containing unreviewed, automatically annotated entries). In release 2010_09 of 10 August 2010, UniProtKB/Swiss-Prot contained 519,348 entries, and UniProtKB/TrEMBL contained 11,636,205 entries.

UniProtKB/Swiss-Prot

UniProtKB/Swiss-Prot is a high-quality, manually annotated, non-redundant protein sequence database. It combines information extracted from scientific literature and biocurator-evaluated computational analysis. The aim of UniProtKB/Swiss-Prot is to provide all known relevant information about a particular protein. Annotation is regularly reviewed to keep up with current scientific findings. The manual annotation of an entry involves detailed analysis of the protein sequence and of the scientific literature.

Sequences from the same gene and the same species are merged into the same database entry. Differences between sequences are identified, and their cause documented (for example alternative splicing, natural variation, incorrect initiation sites, incorrect exon boundaries, frameshifts, unidentified conflicts). A range of sequence analysis tools is used in the annotation of UniProtKB/Swiss-Prot entries. Computer-predictions are manually evaluated, and relevant results selected for inclusion in the entry. These predictions include post-translational modifications, transmembrane domains and topology, signal peptides, domain identification, and protein family classification.

Relevant publications are identified by searching databases such as PubMed. The full text of each paper is read, and information is extracted and added to the entry. Annotation arising from the scientific literature includes, but is not limited to:

- Protein and gene names
- Function
- Enzyme-specific information such as catalytic activity, cofactors and catalytic residues
- Subcellular location
- Protein-protein interactions
- Pattern of expression
- Locations and roles of significant domains and sites
- Ion-, substrate- and cofactor-binding sites
- Protein variant forms produced by natural genetic variation, RNA editing, alternative splicing, proteolytic processing, and post-translational modification

Annotated entries undergo quality assurance before inclusion into UniProtKB/Swiss-Prot. When new data becomes available, entries are updated.

UniProtKB/TrEMBL

UniProtKB/TrEMBL contains high-quality computationally analyzed records, which are enriched with automatic annotation. It was introduced in response to increased dataflow resulting from genome projects, as the time- and labour-consuming manual annotation process of UniProtKB/Swiss-Prot could not be broadened to include all available protein sequences. The translations of annotated coding sequences in the EMBL-Bank/GenBank/DDBJ nucleotide sequence database are automatically processed and entered in UniProtKB/TrEMBL. UniProtKB/TrEMBL also contains sequences from PDB, and from gene prediction, including Ensembl, RefSeq and CCDS.

UniParc

UniProt Archive (UniParc) is a comprehensive and non-redundant database, which contains all the protein sequences from the main, publicly available protein sequence databases. Proteins may exist in several different source databases, and in multiple copies in the same database. In order to avoid redundancy, UniParc stores each unique sequence only once. Identical sequences are merged, regardless of whether they are from the same or different species. Each sequence is given a stable and unique identifier (UPI), making it possible to identify the same protein from different source databases. UniParc contains only protein sequences, with no annotation. When sequences in the source databases change, these changes are tracked by UniParc and history of all changes is archived.

Source Databases

Currently UniParc contains protein sequences from the following publicly available databases:

- INSDC EMBL-Bank/DDBJ/GenBank nucleotide sequence databases
- Ensembl
- European Patent Office (EPO)
- FlyBase
- H-Invitational Database (H-Inv)
- International Protein Index (IPI)
- Japan Patent Office (JPO)
- Protein Information Resource (PIR-PSD)
- Protein Data Bank (PDB)
- Protein Research Foundation (PRF)
- RefSeq
- Saccharomyces Genome Database (SGD)
- The Arabidopsis Information Resource (TAIR)
- TROME
- US Patent Office (USPTO)
- UniProtKB/Swiss-Prot, UniProtKB/Swiss-Prot protein isoforms, UniProtKB/TrEMBL
- Vertebrate and Genome Annotation Database (VEGA)
- WormBase

UniRef

The UniProt Reference Clusters (UniRef) consist of three databases of clustered sets of protein sequences from UniProtKB and selected UniParc records. The UniRef100 database combines identical sequences and sequence fragments (from any organism) into a single UniRef entry. The sequence of a representative protein, the accession numbers of all the merged entries and links to the corresponding UniProtKB and UniParc records are displayed. UniRef100 sequences are clustered using the CD-HIT algorithm to build UniRef90 and UniRef50. Each cluster is composed of sequences that have at least 90% or 50% sequence identity, respectively, to the longest sequence. Clustering sequences significantly reduces database size, enabling faster sequence searches.

UniMes

The UniProt Metagenomic and Environmental Sequences (UniMES) database is a repository specifically developed for metagenomic and environmental data. The predicted proteins from this dataset are combined with automatic classification by InterPro to enhance the original information with further analysis.

UniProtKB contains protein sequences from known species, data arising from metagenomics studies is from environmental (i.e. uncultured) samples and as such the species may not be known/identified.

UniMES was developed for this data. Data from UniMES is not included in UniProtKB or UniRef, but is included in UniParc. UniMES includes data from the Global Ocean Sampling Expedition (GOS). UniMES is available from the UniProt FTP site

Funding for UniProt

UniProt is funded by grants from the National Human Genome Research Institute, the National Institutes of Health (NIH), the European Commission, the Swiss Federal Government through the Federal Office of Education and Science, NCI-caBIG, and the Department of Defense.

- Protein Information Resource (PIR)
- Swiss-Prot
- Protein Data Bank (PDB)
- National Center for Biotechnology Information (NCBI)

The National Center for Biotechnology Information (NCBI) is part of the United States National Library of Medicine (NLM), a branch of the National Institutes of Health.

The NCBI is located in Bethesda, Maryland (38.994994°N 77.099339° WCoordinates: 38.994994°N 77.099339°W) and was founded in 1988 through legislation sponsored by Senator Claude Pepper.

The NCBI houses genome sequencing data in GenBank and an index of biomedical research articles in PubMed Central and PubMed, as well as other information relevant to biotechnology. All these databases are available online through the Entrez search engine.

NCBI is directed by David Lipman, one of the original authors of the BLAST sequence alignment program and a widely respected figure in Bioinformatics. He also leads an intramural research program, including groups led by Stephen Altschul (another BLAST co-author), David Landsman, and Eugene Koonin (a prolific author on comparative genomics).

GenBank

The NCBI has had responsibility for making available the GenBank DNA sequence database since 1992. GenBank coordinates with individual laboratories and other sequence databases such as those of the European Molecular Biology Laboratory (EMBL) and the DNA Data Bank of Japan (DDBJ).

Since 1992, NCBI has grown to provide other databases in addition to GenBank. NCBI provides Online Mendelian Inheritance in Man, the Molecular Modeling Database (3D protein structures), dbSNP a database of single-nucleotide polymorphisms, the Unique Human Gene Sequence Collection, a Gene Map of the human genome, a Taxonomy Browser, and coordinates with the National Cancer Institute to provide the Cancer Genome Anatomy Project. The NCBI assigns a unique identifier (Taxonomy ID number) to each species of organism. The NCBI has software tools that are available by WWW browsing or by FTP. For example, BLAST is a sequence similarity searching program. BLAST can do sequence comparisons against the GenBank DNA database in less than 15 seconds.

The GenBank sequence database is an open access, annotated collection of all publicly available nucleotide sequences and their protein translations. This database is produced and maintained by the National Center for Biotechnology Information (NCBI) as part of the International Nucleotide Sequence Database Collaboration (INSDC). The National Center for Biotechnology Information is a part of the National Institutes of Health in the United States.

GenBank and its collaborators receive sequences produced in laboratories throughout the world from more than 100,000 distinct organisms. In more than 20 years since its establishment, GenBank has become the most important and most influential database for research in almost all biological fields, whose data were accessed and cited by millions of researchers around the world. GenBank continues to grow at an exponential rate, doubling every 18 months. Release 155, produced in August 2006, contained over 65 billion nucleotide bases in more than 61 million sequences. GenBank is built by direct submissions from individual laboratories, as well as from bulk submissions from large-scale sequencing centers.

Submissions

Only original sequences can be submitted to GenBank. Direct submissions are made to GenBank using BankIt, which is a Web-based form, or the stand-alone submission program, Sequin. Upon receipt of a sequence submission, the GenBank staff examines the originality of the data and assigns an accession number to the sequence and performs quality assurance checks. The submissions are then released to the public database, where the entries are retrievable by Entrez or downloadable by FTP. Bulk submissions of Expressed Sequence Tag (EST), Sequence-tagged site (STS), Genome Survey Sequence (GSS), and High-Throughput Genome Sequence (HTGS) data are most often submitted by large-scale sequencing centers. The GenBank direct submissions group also processes complete microbial genome sequences.

History

Walter Goad of the Theoretical Biology and Biophysics Group at Los Alamos National Laboratory and others established the Los Alamos Sequence Database in 1979, which culminated in 1982 with the creation of the public GenBank. Funding was provided by the National Institutes of Health, the National Science Foundation, the Department of Energy, and the Department of Defense. LANL collaborated on GenBank with the firm Bolt, Beranek, and Newman, and by the end of 1983 more than 2,000 sequences were stored in it.

In the mid 1980s, the Intelligenetics bioinformatics company at Stanford University managed the GenBank project in collaboration with LANL. As one of the earliest bioinformatics community projects on the Internet, the GenBank project started BIOSCI/Bionet news groups for promoting open access communications among bioscientists. During 1989 to 1992, the GenBank project transitioned to the newly created National Center for Biotechnology Information.

Growth

The GenBank release notes for release 162.0 (October 2007) state that "from 1982 to the present, the number of bases in GenBank has doubled approximately every 18 months". The following plot clearly shows the exponential growth (on a semi-log scale such as this, a straight line represents an exponential change). The GenBank database includes additional data sets which are constructed mechanically from the main sequence data collection, and therefore are excluded from this count.

NCBI Bookshelf

The NCBI Bookshelf is a collection of freely available, downloadable, on-line versions of selected biomedical books. As of April 2011, the Bookshelf had 845 titles covering a wide range of topics including molecular biology, biochemistry, cell biology, genetics, microbiology, disease states from a molecular and cellular point of view, research methods, and virology. Some of the books are online versions of previously published books, while others, such as Coffee Break (book), are written and edited by NCBI staff. The Bookshelf is a complement to the Entrez PubMed repository of peer-reviewed publication abstracts in that Bookshelf contents provide established perspectives on evolving areas of study and a context in which many disparate individual pieces of reported research can be organized. t

- Human Protein Reference Database
- Proteomics Identifications Database (PRIDE)
- Proteopedia The collaborative, 3D encyclopedia of proteins and other molecules
- Cardiac Organellar Protein Atlas Knowledgebase (COPaKB)

Psychogenomics/ Behavioural Genetics

Behavioural genetics is the field of study that examines the role of genetics in animal (including human) behaviour. Often associated with the "nature versus nurture" debate, behavioural genetics is highly interdisciplinary, involving contributions from biology, genetics, ethology, psychology, and statistics. Behavioural geneticists study the inheritance of behavioural traits. In humans, this information is often gathered through the use of the twin study or adoption study. In animal studies, breeding, transgenesis, and gene knockout techniques are common; psychiatric genetics is a closely related field.

History

Sir Francis Galton, a nineteenth-century intellectual, is recognized as one of the first behavioural geneticists. Galton, a cousin of Charles Darwin, studied the heritability of human ability, focusing on mental characteristics as well as eminence among close relatives in the English upper-class.

In 1869, Galton published his results in Hereditary Genius. In his work, Galton "introduced multivariate analysis and paved the way towards modern Bayesian statistics" that are used throughout the sciences—launching what has been dubbed the "Statistical Enlightenment".

Behaviour genetics, per-se, gained recognition as a research discipline with the publication in 1960 of the textbook Behavior Genetics by J.L. Fuller and W.R. Thompson.

Underscoring the role of evolution in behavioural genetics, Theodosius Dobzhansky was elected the first president of the Behavior Genetics Association in 1972; the BGA bestows the Dobzhansky Award on researchers for their outstanding contributions to the field. In the early 1970s, Lee Ehrman, a doctoral student of Dobzhansky, wrote seminal papers describing the relationship between genotype frequency and mating success in Drosophila, lending impetus to the pursuit of genetic studies of behaviour in other animals.

Methods

The primary methods of modern human behavioural genetics are twin and adoption studies.

Under each of these umbrella terms, these there are large families of designs, such as discordant twin designs. These techniques have been expanded upon in recent decades to include multivariate genetic analysis, extremes analysis, and more recently genomic techniques including the Genome-wide_association_study, linkage, candidate gene approaches and others. Many of the methods of behavioural genetics are also discussed under heritability.

Notable Behavioural Geneticists

Notable behavioural geneticists include Dorret Boomsma, John DeFries, Lindon Eaves, David Fulker, John Hewitt, Kenneth Kendler, John Loehlin, Nick Martin, Gerald McClearn, Robert Plomin, Theodore Reich, who was a pioneer in psychiatric genetics, Hans van Abeelen, Avshalom Caspi, and Steven G. Vandenberg, the founding editor of the journal Behavior Genetics.

Journals

Behavioural geneticists are active in a variety of scientific disciplines including biology, medicine, pharmacology, psychiatry, and psychology; thus, behavioural-genetic research is published in a variety of scientific journals, including Nature and Science.

Journals that specifically publish research in behavioural genetics include Behavior Genetics, Molccular Psychiatry, Psychiatric Genetics, Twin Research and Human Genetics, Genes, Brain and Behavior, and the Journal of Neurogenetics.

Cell

The cell is the functional basic unit of life. It was discovered by Robert Hooke and is the functional unit of all known living organisms. It is the smallest unit of life that is classified as a living thing, and is often called the building block of life.

Organisms can be classified as unicellular (consisting of a single cell; including most bacteria) or multicellular (including plants and animals). Humans contain about 100 trillion cells; a typical cell size is 10 µm and a typical cell mass is 1 nanogram.

The human cell extrema are: largest - anterior horn in the spinal cord (135 ì vs. 120 ì for the ovum), longest - pseudounipolar cells which reach from extremities, including the toes to the lower brain stem, and smallest - granule cells in the cerebellum, at 4 µ. The largest known cells are unfertilised ostrich egg cells, which weigh 3.3 pounds. In 1835, before the final cell theory was developed, Jan Evangelista Purkynì observed small "granules" while looking at the plant tissue through a microscope.

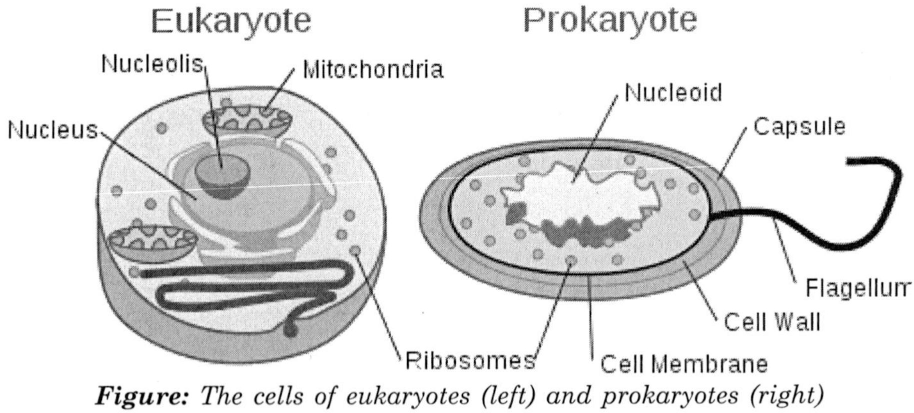

Figure: *The cells of eukaryotes (left) and prokaryotes (right)*

The cell theory, first developed in 1839 by Matthias Jakob Schleiden and Theodor Schwann, states that all organisms are composed of one or more cells, that all cells come from preexisting cells, that vital functions of an organism occur within cells, and that all cells contain the hereditary information necessary for regulating cell functions and for transmitting information to the next generation of cells.

The word cell comes from the Latin cellula, meaning "a small room". The descriptive term for the smallest living biological structure was coined by Robert Hooke in a book he published in 1665 when he compared the cork cells he saw through his microscope to the small rooms monks lived in.

Anatomy

There are two types of cells: eukaryotic and prokaryotic. Prokaryotic cells are usually independent, while eukaryotic cells are often found in multicellular organisms.

Table: *Comparison of features of prokaryotic and eukaryotic cells*

	Prokaryotes	***Eukaryotes***
Typical organisms	*bacteria, archaea*	*protists, fungi, plants, animals*
Typical size	*~ 1-10 μm*	*~ 10-100 μm (sperm cells, apart from the tail, are smaller)*
Type of nucleus	*nucleoid region; no real nucleus*	*real nucleus with double membrane*
DNA	*circular (usually)*	*linear molecules (chromosomes) with histone proteins*
RNA-/protein-synthesis		*coupled in cytoplasm*
RNA-synthesis inside the nucleus		*protein synthesis in cytoplasm*
Ribosomes	*50S+30S*	*60S+40S*
Cytoplasmatic structure	*very few structures*	*highly structured by endomembranes and a cytoskeleton*
Cell movement	*flagella made of flagellin*	*flagella and cilia containing microtubules; lamellipodia and filopodia containing actin*
Mitochondria	*none*	*one to several thousand (though some lack mitochondria)*
Chloroplasts	*none*	*in algae and plants*

Organization	*usually single cells*	*single cells, colonies, higher multicellular organisms with specialized cells*
Cell division	*Binary fission (simple division)*	*Mitosis (fission or budding) Meiosis*

Prokaryotic Cells

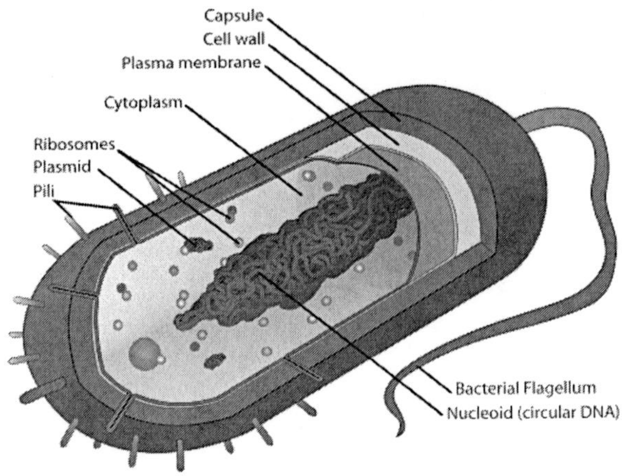

Figure : Diagram of a typical prokaryotic cell

The prokaryote cell is simpler, and therefore smaller, than a eukaryote cell, lacking a nucleus and most of the other organelles of eukaryotes. There are two kinds of prokaryotes: bacteria and archaea; these share a similar structure. Nuclear material of prokaryotic cell consist of a single chromosome that is in direct contact with cytoplasm. Here, the undefined nuclear region in the cytoplasm is called nucleoid.

A prokaryotic cell has three architectural regions:

- On the outside, flagella and pili project from the cell's surface. These are structures (not present in all prokaryotes) made of proteins that facilitate movement and communication between cells;
- Enclosing the cell is the cell envelope – generally consisting of a cell wall covering a plasma membrane though some bacteria also have a further covering layer called a capsule. The envelope gives rigidity to the cell and separates the interior of the cell from its environment, serving as a protective filter. Though most prokaryotes have a cell wall, there are exceptions such as Mycoplasma (bacteria) and Thermoplasma (archaea). The cell

wall consists of peptidoglycan in bacteria, and acts as an additional barrier against exterior forces. It also prevents the cell from expanding and finally bursting (cytolysis) from osmotic pressure against a hypotonic environment. Some eukaryote cells (plant cells and fungi cells) also have a cell wall;

- Inside the cell is the cytoplasmic region that contains the cell genome (DNA) and ribosomes and various sorts of inclusions. A prokaryotic chromosome is usually a circular molecule (an exception is that of the bacterium Borrelia burgdorferi, which causes Lyme disease). Though not forming a nucleus, the DNA is condensed in a nucleoid. Prokaryotes can carry extrachromosomal DNA elements called plasmids, which are usually circular. Plasmids enable additional functions, such as antibiotic resistance.

Eukaryotic Cells

Plants, animals, fungi, slime moulds, protozoa, & algae are all Eukaryotic. These cells are about 15 times wider than a typical prokaryote and can be as much as 1000 times greater in volume. The major difference between prokaryotes and eukaryotes is that eukaryotic cells contain membrane-bound compartments in which specific metabolic activities take place. Most important among these is a cell nucleus, a membrane-delineated compartment that houses the eukaryotic cell's DNA. This nucleus gives the eukaryote its name, which means "true nucleus." Other differences include:

- The plasma membrane resembles that of prokaryotes in function, with minor differences in the setup. Cell walls may or may not be present.
- The eukaryotic DNA is organized in one or more linear molecules, called chromosomes, which are associated with histone proteins. All chromosomal DNA is stored in the cell nucleus, separated from the cytoplasm by a membrane. Some eukaryotic organelles such as mitochondria also contain some DNA.
- Many eukaryotic cells are ciliated with primary cilia. Primary cilia play important roles in chemosensation, mechanosensation, and thermosensation. Cilia may thus be "viewed as sensory cellular antennae that coordinate a large number of cellular signaling pathways, sometimes coupling the signaling to ciliary motility or alternatively to cell division and differentiation."
- Eukaryotes can move using motile cilia or flagella. The flagella are more complex than those of prokaryotes.

Subcellular Components

All cells, whether prokaryotic or eukaryotic, have a membrane that envelops the cell, separates its interior from its environment, regulates what moves in and out (selectively permeable), and maintains the electric potential of the cell. Inside the membrane, a salty cytoplasm takes up most of the cell volume.

All cells possess DNA, the hereditary material of genes, and RNA, containing the information necessary to build various proteins such as enzymes, the cell's primary machinery. There are also other kinds of biomolecules in cells. This article lists these primary components of the cell, then briefly describe their function.

Membrane

The cytoplasm of a cell is surrounded by a cell membrane or plasma membrane. The plasma membrane in plants and prokaryotes is usually covered by a cell wall. This membrane serves to separate and protect a cell from its surrounding environment and is made mostly from a double layer of lipids (hydrophobic fat-like molecules) and hydrophilic phosphorus molecules. Hence, the layer is called a phospholipid bilayer.

It may also be called a fluid mosaic membrane. Embedded within this membrane is a variety of protein molecules that act as channels and pumps that move different molecules into and out of the cell. The membrane is said to be 'semi-permeable', in that it can either let a substance (molecule or ion) pass through freely, pass through to a limited extent or not pass through at all. Cell surface membranes also contain receptor proteins that allow cells to detect external signaling molecules such as hormones.

Cytoskeleton

The cytoskeleton acts to organize and maintain the cell's shape; anchors organelles in place; helps during endocytosis, the uptake of external materials by a cell, and cytokinesis, the separation of daughter cells after cell division; and moves parts of the cell in processes of growth and mobility.

The eukaryotic cytoskeleton is composed of microfilaments, intermediate filaments and microtubules. There is a great number of proteins associated with them, each controlling a cell's structure by directing, bundling, and aligning filaments. The prokaryotic cytoskeleton is less well-studied but is involved in the maintenance of cell shape, polarity and cytokinesis.

Genetic Material

Two different kinds of genetic material exist: deoxyribonucleic acid (DNA) and ribonucleic acid (RNA). Most organisms use DNA for their long-term information storage, but some viruses (e.g., retroviruses) have RNA as their genetic material. The biological information contained in an organism is encoded in its DNA or RNA sequence. RNA is also used for information transport (e.g., mRNA) and enzymatic functions (e.g., ribosomal RNA) in organisms that use DNA for the genetic code itself. Transfer RNA (tRNA) molecules are used to add amino acids during protein translation.

Prokaryotic genetic material is organized in a simple circular DNA molecule (the bacterial chromosome) in the nucleoid region of the cytoplasm. Eukaryotic genetic material is divided into different, linear molecules called chromosomes inside a discrete nucleus, usually with additional genetic material in some organelles like mitochondria and chloroplasts. A human cell has genetic material contained in the cell nucleus (the nuclear genome) and in the mitochondria (the mitochondrial genome). In humans the nuclear genome is divided into 23 pairs of linear DNA molecules called chromosomes. The mitochondrial genome is a circular DNA molecule distinct from the nuclear DNA. Although the mitochondrial DNA is very small compared to nuclear chromosomes, it codes for 13 proteins involved in mitochondrial energy production and specific tRNAs.

Foreign genetic material (most commonly DNA) can also be artificially introduced into the cell by a process called transfection. This can be transient, if the DNA is not inserted into the cell's genome, or stable, if it is. Certain viruses also insert their genetic material into the genome.

Organelles

The human body contains many different organs, such as the heart, lung, and kidney, with each organ performing a different function. Cells also have a set of "little organs," called organelles, that are adapted and/or specialized for carrying out one or more vital functions. Both eukaryotic and prokaryotic cells have organelles but organelles in eukaryotes are generally more complex and may be membrane bound.

There are several types of organelles in a cell. Some (such as the nucleus and golgi apparatus) are typically solitary, while others (such as mitochondria, peroxisomes and lysosomes) can be numerous (hundreds to thousands). The cytosol is the gelatinous fluid that fills the cell and surrounds the organelles.

Cell Nucleus - Eukaryotes only - A Cell's Information Center

The cell nucleus is the most conspicuous organelle found in a eukaryotic cell. It houses the cell's chromosomes, and is the place where almost all DNA replication and RNA synthesis (transcription) occur. The nucleus is spherical and separated from the cytoplasm by a double membrane called the nuclear envelope. The nuclear envelope isolates and protects a cell's DNA from various molecules that could accidentally damage its structure or interfere with its processing. During processing, DNA is transcribed, or copied into a special RNA, called messenger RNA (mRNA). This mRNA is then transported out of the nucleus, where it is translated into a specific protein molecule. The nucleolus is a specialized region within the nucleus where ribosome subunits are assembled. In prokaryotes, DNA processing takes place in the cytoplasm.

Mitochondria and Chloroplasts - Eukaryotes only - the Power Generators

Mitochondria are self-replicating organelles that occur in various numbers, shapes, and sizes in the cytoplasm of all eukaryotic cells. Mitochondria play a critical role in generating energy in the eukaryotic cell. Mitochondria generate the cell's energy by oxidative phosphorylation, using oxygen to release energy stored in cellular nutrients (typically pertaining to glucose) to generate ATP. Mitochondria multiply by splitting in two. Respiration occurs in the cell mitochondria.

Organelles that are modified chloroplasts are broadly called plastids, and are involved in energy storage through photosynthesis, which uses solar energy to generate carbohydrates and oxygen from carbon dioxide and water.

Mitochondria and chloroplasts each contain their own genome, which is separate and distinct from the nuclear genome of a cell. Both organelles contain this DNA in circular plasmids, much like prokaryotic cells, strongly supporting the evolutionary theory of endosymbiosis; since these organelles contain their own genomes and have other similarities to prokaryotes, they are thought to have developed through a symbiotic relationship after being engulfed by a primitive cell.

Endoplasmic Reticulum - Eukaryotes only

The endoplasmic reticulum (ER) is the transport network for molecules targeted for certain modifications and specific destinations, as compared to molecules that float freely in the cytoplasm. The ER has two forms: the rough ER, which has ribosomes on its surface and secretes proteins into the cytoplasm, and the smooth ER, which lacks them. Smooth ER plays a role in calcium sequestration and release.

Golgi Apparatus - Eukaryotes only

The primary function of the Golgi apparatus is to process and package the macromolecules such as proteins and lipids that are synthesized by the cell. It is particularly important in the processing of proteins for secretion. The Golgi apparatus forms a part of the endomembrane system of eukaryotic cells. Vesicles that enter the Golgi apparatus are processed in a cis to trans direction, meaning they coalesce on the cis side of the apparatus and after processing pinch off on the opposite (trans) side to form a new vesicle in the animal cell.

Ribosomes

The ribosome is a large complex of RNA and protein molecules. They each consist of two subunits, and act as an assembly line where RNA from the nucleus is used to synthesise proteins from amino acids. Ribosomes can be found either floating freely or bound to a membrane (the rough endoplasmatic reticulum in eukaryotes, or the cell membrane in prokaryotes).

Lysosomes and Peroxisomes - Eukaryotes only

Lysosomes contain digestive enzymes (acid hydrolases). They digest excess or worn-out organelles, food particles, and engulfed viruses or bacteria. Peroxisomes have enzymes that rid the cell of toxic peroxides. The cell could not house these destructive enzymes if they were not contained in a membrane-bound system. These organelles are often called a "suicide bag" because of their ability to detonate and destroy the cell.

Centrosome - the Cytoskeleton Organiser

The centrosome produces the microtubules of a cell - a key component of the cytoskeleton. It directs the transport through the ER and the Golgi apparatus. Centrosomes are composed of two centrioles, which separate during cell division and help in the formation of the mitotic spindle. A single centrosome is present in the animal cells. They are also found in some fungi and algae cells.

Vacuoles

Vacuoles store food and waste. Some vacuoles store extra water. They are often described as liquid filled space and are surrounded by a membrane. Some cells, most notably Amoeba, have contractile vacuoles, which can pump water out of the cell if there is too much water. The vacuoles of eukaryotic cells are usually larger in those of plants than animals.

Structures Outside the Cell Wall

Capsule

A gelatinous capsule is present in some bacteria outside the cell wall. The capsule may be polysaccharide as in pneumococci, meningococci or polypeptide as Bacillus anthracis or hyaluronic acid as in streptococci. Capsules are not marked by ordinary stain and can be detected by special stain. The capsule is antigenic. The capsule has antiphagocytic function so it determines the virulence of many bacteria. It also plays a role in attachment of the organism to mucous membranes.

Flagella

Flagella are the organelles of cellular mobility. They arise from cytoplasm and extrude through the cell wall. They are long and thick thread-like appendages, protein in nature. Are most commonly found in bacteria cells but are found in animal cells as well.

Fimbriae (pili)

They are short and thin hair like filaments, formed of protein called pilin (antigenic). Fimbriae are responsible for attachment of bacteria to specific receptors of human cell (adherence). There are special types of pili called (sex pili) involved in conjunction.

The Basic Microarray

Since an array can contain tens of thousands of probes, a microarray experiment can accomplish many genetic tests in parallel. Therefore arrays have dramatically accelerated many types of investigation. In standard microarrays, the probes are synthesized and then attached via surface engineering to a solid surface by a covalent bond to a chemical matrix (via epoxy-silane, amino-silane, lysine, polyacrylamide or others).

The solid surface can be glass or a silicon chip, in which case they are colloquially known as an Affy chip when an Affymetrix chip is used. Other microarray platforms, such as Illumina, use microscopic beads, instead of the large solid support. Alternatively, microarrays can be constructed by the direct synthesis of oligonucleotide probes on solid surfaces. DNA arrays are different from other types of microarray only in that they either measure DNA or use DNA as part of its detection system.

DNA microarrays can be used to measure changes in expression levels, to detect single nucleotide polymorphisms (SNPs), or to genotype or resequence mutant genomes. Microarrays also differ in fabrication, workings, accuracy, efficiency, and cost. Additional factors for microarray experiments are the experimental design and the methods of analyzing the data.

History

Microarray technology evolved from Southern blotting, where fragmented DNA is attached to a substrate and then probed with a known gene or fragment.

The first reported use of this approach was the analysis of 378 arrayed lysed bacterial colonies each harboring a different sequence which were assayed in multiple replicas for expression of the genes in multiple normal and tumor tissue.

This was expanded to analysis of more than 4000 human sequences with computer driven scanning and image processing for quantitative analysis of the sequences in human colonic tumors and normal tissue and then to comparison of colonic tissues at different genetic risk. The use of a collection of distinct DNAs in arrays for expression profiling was also described in 1987, and the arrayed DNAs were used to identify genes whose expression is modulated by interferon.

These early gene arrays were made by spotting cDNAs onto filter paper with a pin-spotting device. The use of miniaturized microarrays for gene expression profiling was first reported in 1995, and a complete eukaryotic genome (Saccharomyces cerevisiae) on a microarray was published in 1997.

Principle

Hybridization of the Target to the Probe

The core principle behind microarrays is hybridization between two DNA strands, the property of complementary nucleic acid sequences to specifically pair with each other by forming hydrogen bonds between complementary nucleotide base pairs. A high number of complementary base pairs in a nucleotide sequence means tighter non-covalent bonding between the two strands.

After washing off of non-specific bonding sequences, only strongly paired strands will remain hybridized. So fluorescently labeled target sequences that bind to a probe sequence generate a signal that depends on the strength of the hybridization determined by the number of paired bases, the hybridization conditions (such as temperature), and washing after hybridization. Total strength of the signal, from a spot (feature), depends upon the amount of target sample binding to the probes present on that spot. Microarrays use relative quantization in which the intensity of a feature is compared to the intensity of the same feature under a different condition, and the identity of the feature is known by its position.

Uses and Types

Many types of array exist and the broadest distinction is whether they are spatially arranged on a surface or on coded beads:

- The traditional solid-phase array is a collection of orderly microscopic "spots", called features, each with a specific probe attached to a solid surface, such as glass, plastic or silicon biochip (commonly known as a genome chip, DNA chip or gene array). Thousands of them can be placed in known locations on a single DNA microarray.
- The alternative bead array is a collection of microscopic polystyrene beads, each with a specific probe and a ratio of two or more dyes, which do not interfere with the fluorescent dyes used on the target sequence.

DNA microarrays can be used to detect DNA (as in comparative genomic hybridization), or detect RNA (most commonly as cDNA after reverse transcription) that may or may not be translated into proteins.

The process of measuring gene expression via cDNA is called expression analysis or expression profiling.

Applications include

Application or Technology	*Synopsis*
Gene expression profiling	In an mRNA or gene expression profiling experiment the expression levels of thousands of genes are simultaneously monitored to study the effects of certain treatments, diseases, and developmental stages on gene expression. For example, microarray-based gene expression profiling can be used to identify genes whose expression is changed in response to pathogens or other organisms by comparing gene expression in infected to that in uninfected cells or tissues.
Comparative genomic hybridization	Assessing genome content in different cells or closely related organisms.
GeneID	Small microarrays to check IDs of organisms in food and feed (like GMO), mycoplasms in cell culture, or pathogens

	for disease detection, mostly combining PCR and microarray technology.
Chromatin immunoprecipitation on Chip	DNA sequences bound to a particular protein can be isolated by immunoprecipitating that protein (ChIP), these fragments can be then hybridized to a microarray (such as a tiling array) allowing the determination of protein binding site occupancy throughout the genome. Example protein to immunoprecipitate are histone modifications (H3K27me3, H3K4me2, H3K9me3, etc.), Polycomb-group protein (PRC2:Suz12, PRC1:YY1) and trithorax-group protein (Ash1) to study the epigenetic landscape or RNA Polymerase II to study the transcription landscape.
DamID	Analogously to ChIP, genomic regions bound by a protein of interest can be isolated and used to probe a microarray to determine binding site occupancy. Unlike ChIP, DamID does not require antibodies but makes use of adenine methylation near the protein's binding sites to selectively amplify those regions, introduced by expressing minute amounts of protein of interest fused to bacterial DNA adenine methyltransferase.
SNP detection	Identifying single nucleotide polymorphism among alleles within or between populations. Several applications of microarrays make use of SNP detection, including Genotyping, forensic analysis, measuring predisposition to disease, identifying drug-candidates, evaluating germline mutations in individuals or somatic mutations in cancers, assessing loss of heterozygosity, or genetic linkage analysis.
Alternative splicing detection	An 'exon junction array design uses probes specific to the expected or potential splice sites of predicted exons for a gene. It is of intermediate density, or coverage, to a typical gene expression array (with 1-3 probes per gene) and a genomic tiling array (with hundreds or thousands of probes per gene). It is used to assay the expression of alternative splice forms of a gene. Exon arrays have a different

	design, employing probes designed to detect each individual exon for known or predicted genes, and can be used for detecting different splicing isoforms.
Fusion genes microarray	A Fusion gene microarray can detect fusion transcripts, e.g. from cancer specimens. The principle behind this is building on the alternative splicing microarrays. The oligo design strategy enables combined measurements of chimeric transcript junctions with exon-wise measurements of individual fusion partners.
Tiling array	Genome tiling arrays consist of overlapping probes designed to densely represent a genomic region of interest, sometimes as large as an entire human chromosome. The purpose is to empirically detect expression of transcripts or alternatively splice forms which may not have been previously known or predicted.

Fabrication

Microarrays can be manufactured in different ways, depending on the number of probes under examination, costs, customization requirements, and the type of scientific question being asked. Arrays may have as few as 10 probes or up to 2.1 million micrometre-scale probes from commercial vendors.

Spotted vs. in Situ Synthesised Arrays

Microarrays can be fabricated using a variety of technologies, including printing with fine-pointed pins onto glass slides, photolithography using pre-made masks, photolithography using dynamic micromirror devices, ink-jet printing, or electrochemistry on microelectrode arrays.

In spotted microarrays, the probes are oligonucleotides, cDNA or small fragments of PCR products that correspond to mRNAs. The probes are synthesized prior to deposition on the array surface and are then "spotted" onto glass. A common approach utilizes an array of fine pins or needles controlled by a robotic arm that is dipped into wells containing DNA probes and then depositing each probe at designated locations on the array surface. The resulting "grid" of probes represents the nucleic acid profiles of the prepared probes and is ready to receive complementary cDNA or cRNA "targets" derived from experimental or clinical samples. This technique is used by research scientists around the world to produce "in-house" printed microarrays from their own labs. These arrays may be easily customized for each experiment, because researchers can choose the probes and printing locations on the arrays, synthesize the probes in their own lab (or collaborating facility), and spot the arrays.

They can then generate their own labeled samples for hybridization, hybridize the samples to the array, and finally scan the arrays with their own equipment.

This provides a relatively low-cost microarray that may be customized for each study, and avoids the costs of purchasing often more expensive commercial arrays that may represent vast numbers of genes that are not of interest to the investigator.

Publications exist which indicate in-house spotted microarrays may not provide the same level of sensitivity compared to commercial oligonucleotide arrays, possibly owing to the small batch sizes and reduced printing efficiencies when compared to industrial manufactures of oligo arrays.

In oligonucleotide microarrays, the probes are short sequences designed to match parts of the sequence of known or predicted open reading frames. Although oligonucleotide probes are often used in "spotted" microarrays, the term "oligonucleotide array" most often refers to a specific technique of manufacturing. Oligonucleotide arrays are produced by printing short oligonucleotide sequences designed to represent a single gene or family of gene splice-variants by synthesizing this sequence directly onto the array surface instead of depositing intact sequences. Sequences may be longer (60-mer probes such as the Agilent design) or shorter (25-mer probes produced by Affymetrix) depending on the desired purpose; longer probes are more specific to individual target genes, shorter probes may be spotted in higher density across the array and are cheaper to manufacture.

One technique used to produce oligonucleotide arrays include photolithographic synthesis (Affymetrix) on a silica substrate where light and light-sensitive masking agents are used to "build" a sequence one nucleotide at a time across the entire array. Each applicable probe is selectively "unmasked" prior to bathing the array in a solution of a single nucleotide, then a masking reaction takes place and the next set of probes are unmasked in preparation for a different nucleotide exposure. After many repetitions, the sequences of every probe become fully constructed. More recently, Maskless Array Synthesis from NimbleGen Systems has combined flexibility with large numbers of probes.

Two-channel vs. One-channel Detection

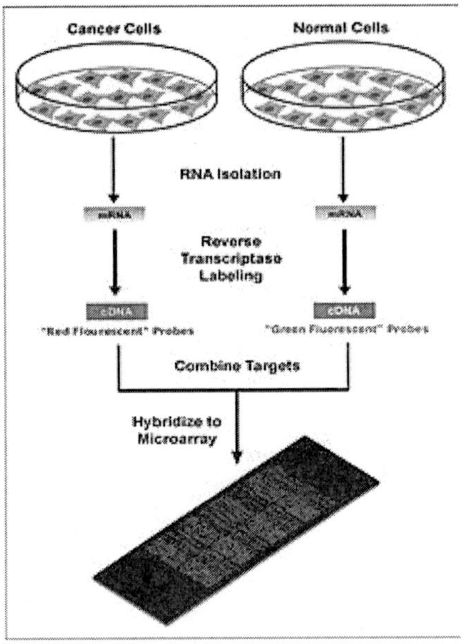

Figure: Diagram of typical dual-colour microarray experiment.

Two-color microarrays or two-channel microarrays are typically hybridized with cDNA prepared from two samples to be compared (e.g. diseased tissue versus healthy tissue) and that are labeled with two different fluorophores. Fluorescent dyes commonly used for cDNA labeling include Cy3, which has a fluorescence emission wavelength of 570 nm (corresponding to the green part of the light spectrum), and Cy5 with a fluorescence emission wavelength of 670 nm (corresponding to the red part of the light spectrum). The two Cy-labeled cDNA samples are mixed and hybridized to a single microarray that is then scanned in a microarray scanner to visualize fluorescence of the two fluorophores after excitation with a laser beam of a defined wavelength. Relative intensities of each fluorophore may then be used in ratio-based analysis to identify up-regulated and down-regulated genes.

Oligonucleotide microarrays often carry control probes designed to hybridize with RNA spike-ins. The degree of hybridization between the spike-ins and the control probes is used to normalize the hybridization measurements for the target probes. Although absolute levels of gene expression may be determined in the two-color array in rare instances, the relative differences in expression among different spots within a sample and between samples is the preferred method of data analysis for the two-color system. Examples of providers for such microarrays includes Agilent with their Dual-Mode platform,

Eppendorf with their DualChip platform for colorimetric Silverquant labeling, and TeleChem International with Arrayit. In single-channel microarrays or one-color microarrays, the arrays provide intensity data for each probe or probe set indicating a relative level of hybridization with the labeled target.

However, they do not truly indicate abundance levels of a gene but rather relative abundance when compared to other samples or conditions when processed in the same experiment. Each RNA molecule encounters protocol and batch-specific bias during amplification, labeling, and hybridization phases of the experiment making comparisons between genes for the same microarray uninformative. The comparison of two conditions for the same gene requires two separate single-dye hybridizations. Several popular single-channel systems are the Affymetrix "Gene Chip", Illumina "Bead Chip", Agilent single-channel arrays, the Applied Microarrays "CodeLink" arrays, and the Eppendorf "DualChip & Silverquant". One strength of the single-dye system lies in the fact that an aberrant sample cannot affect the raw data derived from other samples, because each array chip is exposed to only one sample (as opposed to a two-color system in which a single low-quality sample may drastically impinge on overall data precision even if the other sample was of high quality). Another benefit is that data are more easily compared to arrays from different experiments so long as batch effects have been accounted for. A drawback to the one-color system is that, when compared to the two-color system, twice as many microarrays are needed to compare samples within an experiment.

Microarrays and Bioinformatics

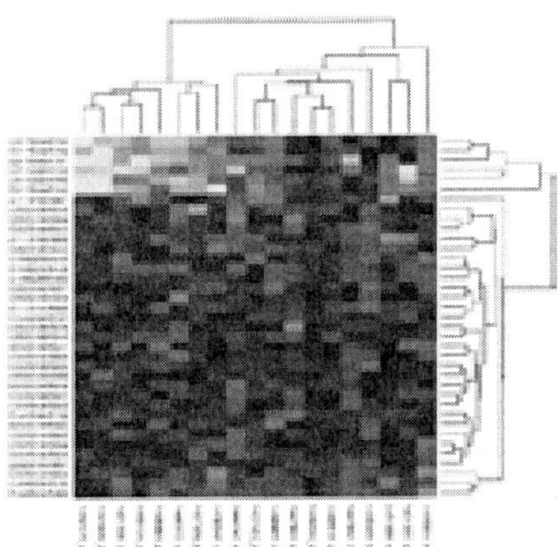

Figure:

Gene expression values from microarray experiments can be represented as heat maps to visualize the result of data analysis. The advent of inexpensive microarray experiments created several specific bioinformatics challenges:

- The multiple levels of replication in experimental design (Experimental design)
- The number of platforms and independent groups and data format (Standardization)
- The treatment of the data (Statistical analysis)
- Accuracy and precision (Relation between probe and gene)
- The sheer volume of data and the ability to share it (Data warehousing)

Experimental Design

Due to the biological complexity of gene expression, the considerations of experimental design that are discussed in the expression profiling article are of critical importance if statistically and biologically valid conclusions are to be drawn from the data.

There are three main elements to consider when designing a microarray experiment. First, replication of the biological samples is essential for drawing conclusions from the experiment. Second, technical replicates (two RNA samples obtained from each experimental unit) help to ensure precision and allow for testing differences within treatment groups. The biological replicates include independent RNA extractions and technical replicates may be two aliquots of the same extraction. Third, spots of each cDNA clone or oligonucleotide are present as replicates (at least duplicates) on the microarray slide, to provide a measure of technical precision in each hybridization. It is critical that information about the sample preparation and handling is discussed, in order to help identify the independent units in the experiment and to avoid inflated estimates of statistical significance.

Standardization

Microarray data is difficult to exchange due to the lack of standardization in platform fabrication, assay protocols, and analysis methods. This presents an interoperability problem in bioinformatics. Various grass-roots open-source projects are trying to ease the exchange and analysis of data produced with non-proprietary chips:

- For example, the "Minimum Information About a Microarray Experiment" (MIAME) checklist helps define the level of detail that should exist and is being adopted by many journals as a requirement for the submission of papers incorporating microarray results. But MIAME does not describe the format for the information, so while many formats can support the MIAME requirements, as of 2007 no format permits verification of complete semantic compliance.

- The "MicroArray Quality Control (MAQC) Project" is being conducted by the US Food and Drug Administration (FDA) to develop standards and quality control metrics which will eventually allow the use of MicroArray data in drug discovery, clinical practice and regulatory decision-making.
- The MGED Society has developed standards for the representation of gene expression experiment results and relevant annotations.

Statistical Analysis

Microarray data sets are commonly very large, and analytical precision is influenced by a number of variables. Statistical challenges include taking into account effects of background noise and appropriate normalization of the data.

Normalization methods may be suited to specific platforms and, in the case of commercial platforms, the analysis may be proprietary. Algorithms that affect statistical analysis include:

- Image analysis: gridding, spot recognition of the scanned image (segmentation algorithm), removal or marking of poor-quality and low-intensity features (called flagging).
- Data processing: background subtraction (based on global or local background), determination of spot intensities and intensity ratios, visualisation of data, and log-transformation of ratios, global or local normalization of intensity ratios, and segmentation into different copy number regions using step detection algorithms.
- Identification of statistically significant changes: t-test, ANOVA, Bayesian method Mann–Whitney test methods tailored to microarray data sets, which take into account multiple comparisons or cluster analysis. These methods assess statistical power based on the variation present in the data and the number of experimental replicates, and can help minimize Type I and type II errors in the analyses.
- Network-based methods: Statistical methods that take the underlying structure of gene networks into account, representing either associative or causative interactions or dependencies among gene products.

Microarray data may require further processing aimed at reducing the dimensionality of the data to aid comprehension and more focused analysis. Other methods permit analysis of data consisting of a low number of biological or technical replicates; for example, the Local Pooled Error (LPE) test pools standard deviations of genes with similar expression levels in an effort to compensate for insufficient replication.

Relation Between Probe and Gene

The relation between a probe and the mRNA that it is expected to detect is not trivial. Some mRNAs may cross-hybridize probes in the array that are supposed to detect another mRNA.

In addition, mRNAs may experience amplification bias that is sequence or molecule-specific. Thirdly, probes that are designed to detect the mRNA of a particular gene may be relying on genomic EST information that is incorrectly associated with that gene.

Data Warehousing

Microarray data was found to be more useful when compared to other similar datasets. The sheer volume of data, specialized formats (such as MIAME), and curation efforts associated with the datasets require specialized databases to store the data.

DNA sequencing

DNA sequencing includes several methods and technologies that are used for determining the order of the nucleotide bases—adenine, guanine, cytosine, and thymine—in a molecule of DNA.

Knowledge of DNA sequences has become indispensable for basic biological research, other research branches utilizing DNA sequencing, and in numerous applied fields such as diagnostic, biotechnology, forensic biology and biological systematics.

The advent of DNA sequencing has significantly accelerated biological research and discovery. The rapid speed of sequencing attained with modern DNA sequencing technology has been instrumental in the sequencing of the human genome, in the Human Genome Project. Related projects, often by scientific collaboration across continents, have generated the complete DNA sequences of many animal, plant, and microbial genomes.

The first DNA sequences were obtained in the early 1970s by academic researchers using laborious methods based on two-dimensional chromatography. Following the development of dye-based sequencing methods with automated analysis, DNA sequencing has become easier and orders of magnitude faster.

History

RNA sequencing was one of the earliest forms of nucleotide sequencing. The major landmark of RNA sequencing is the sequence of the first complete gene and the complete genome of Bacteriophage MS2, identified and published by Walter Fiers and his coworkers at the University of Ghent (Ghent, Belgium), between 1972 and 1976.

Prior to the development of rapid DNA sequencing methods in the early 1970s by Frederick Sanger at the University of Cambridge, in England and Walter Gilbert and Allan Maxam at Harvard, a number of laborious methods were used. For instance, in 1973, Gilbert and Maxam reported the sequence of 24 basepairs using a method known as wandering-spot analysis.

The chain-termination method developed by Sanger and coworkers in 1977 soon became the method of choice, owing to its relative ease and reliability. It involves separating DNA bases from different DNA fragments.

Maxam–Gilbert Sequencing

In 1976–1977, Allan Maxam and Walter Gilbert developed a DNA sequencing method based on chemical modification of DNA and subsequent cleavage at specific bases. Although Maxam and Gilbert published their chemical sequencing method two years after the ground-breaking paper of Sanger and Coulson on plus-minus sequencing, Maxam–Gilbert sequencing rapidly became more popular, since purified DNA could be used directly, while the initial Sanger method required that each read start be cloned for production of single-stranded DNA. However, with the improvement of the chain-termination method, Maxam-Gilbert sequencing has fallen out of favour due to its technical complexity prohibiting its use in standard molecular biology kits, extensive use of hazardous chemicals, and difficulties with scale-up.

The method requires radioactive labeling at one 5' end of the DNA (typically by a kinase reaction using gamma- P ATP) and purification of the DNA fragment to be sequenced. Chemical treatment generates breaks at a small proportion of one or two of the four nucleotide bases in each of four reactions (G, A+G, C, C+T). For example, the purines (A+G) are depurinated using formic acid, the guanines (and to some extent the adenines) are methylated by dimethyl sulfate, and the pyrimidines (C+T) are methylated using hydrazine. The addition of salt (sodium chloride) to the hydrazine reaction inhibits the methylation of thymine for the C-only reaction.

The modified DNAs are then cleaved by hot piperidine at the position of the modified base. The concentration of the modifying chemicals is controlled to introduce on average one modification per DNA molecule.

Thus a series of labeled fragments is generated, from the radiolabeled end to the first "cut" site in each molecule. The fragments in the four reactions are electrophoresed side by side in denaturing acrylamide gels for size separation. To visualize the fragments, the gel is exposed to X-ray film for autoradiography, yielding a series of dark bands each corresponding to a radiolabeled DNA fragment, from which the sequence may be inferred. Also sometimes known as "chemical sequencing", this method led to the Methylation Interference Assay used to map DNA-binding sites for DNA-binding proteins.

Chain-termination Methods

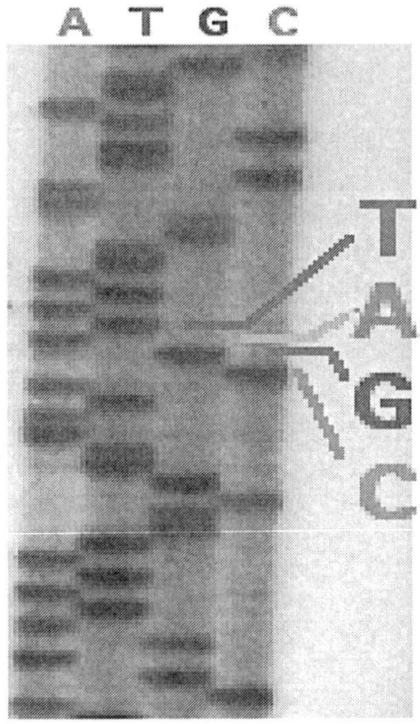

Figure: Part of a radioactively labelled sequencing gel

Because the chain-terminator method (or Sanger method after its developer Frederick Sanger) is more efficient and uses fewer toxic chemicals and lower amounts of radioactivity than the method of Maxam and Gilbert, it rapidly became the method of choice. The key principle of the Sanger method was the use of dideoxynucleotide triphosphates (ddNTPs) as DNA chain terminators. The classical chain-termination method requires a single-stranded DNA template, a DNA primer, a DNA polymerase, normal deoxynucleotide triphosphates (dNTPs), and modified nucleotides (dideoxyNTPs) that terminate DNA strand elongation. These ddNTPs will also be radioactively or fluorescently labelled for detection in automated sequencing machines. The DNA sample is divided into four separate sequencing reactions, containing all four of the standard deoxynucleotides (dATP, dGTP, dCTP and dTTP) and the DNA polymerase. To each reaction is added only one of the four dideoxynucleotides (ddATP, ddGTP, ddCTP, or ddTTP) which are the chain-terminating nucleotides, lacking a 3'-OH group required for the formation of a phosphodiester bond between two nucleotides, thus terminating DNA strand extension and resulting in DNA fragments of varying length.

The newly synthesized and labelled DNA fragments are heat denatured, and separated by size (with a resolution of just one nucleotide) by gel electrophoresis on a denaturing polyacrylamide-urea gel with each of the four reactions run in one of four individual lanes (lanes A, T, G, C); the DNA bands are then visualized by autoradiography or UV light, and the DNA sequence can be directly read off the X-ray film or gel image. In the image on the right, X-ray film was exposed to the gel, and the dark bands correspond to DNA fragments of different lengths. A dark band in a lane indicates a DNA fragment that is the result of chain termination after incorporation of a dideoxynucleotide (ddATP, ddGTP, ddCTP, or ddTTP). The relative positions of the different bands among the four lanes are then used to read (from bottom to top) the DNA sequence. DNA fragments are labelled with a radioactive or fluorescent tag on the primer (1), in the new DNA strand with a labeled dNTP, or with a labeled ddNTP. (click to expand)

Technical variations of chain-termination sequencing include tagging with nucleotides containing radioactive phosphorus for radiolabelling, or using a primer labeled at the 5' end with a fluorescent dye. Dye-primer sequencing facilitates reading in an optical system for faster and more economical analysis and automation. The later development by Leroy Hood and coworkers of fluorescently labeled ddNTPs and primers set the stage for automated, high-throughput DNA sequencing. Sequence ladder by radioactive sequencing compared to fluorescent peaks (click to expand) Chain-termination methods have greatly simplified DNA sequencing. For example, chain-termination-based kits are commercially available that contain the reagents needed for sequencing, pre-aliquoted and ready to use. Limitations include non-specific binding of the primer to the DNA, affecting accurate read-out of the DNA sequence, and DNA secondary structures affecting the fidelity of the sequence.

Dye-terminator Sequencing

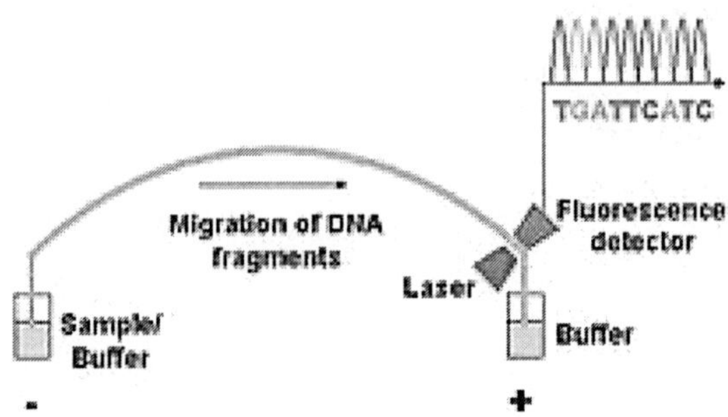

Figure: Capillary electrophoresis (click to expand)

Dye-terminator sequencing utilizes labelling of the chain terminator ddNTPs, which permits sequencing in a single reaction, rather than four reactions as in the labelled-primer method. In dye-terminator sequencing, each of the four dideoxynucleotide chain terminators is labelled with fluorescent dyes, each of which emit light at different wavelengths. Owing to its greater expediency and speed, dye-terminator sequencing is now the mainstay in automated sequencing. Its limitations include dye effects due to differences in the incorporation of the dye-labelled chain terminators into the DNA fragment, resulting in unequal peak heights and shapes in the electronic DNA sequence trace chromatogram after capillary electrophoresis. This problem has been addressed with the use of modified DNA polymerase enzyme systems and dyes that minimize incorporation variability, as well as methods for eliminating "dye blobs". The dye-terminator sequencing method, along with automated high-throughput DNA sequence analyzers, is now being used for the vast majority of sequencing projects.

Challenges

Common challenges of DNA sequencing include poor quality in the first 15–40 bases of the sequence and deteriorating quality of sequencing traces after 700–900 bases. Base calling software typically gives an estimate of quality to aid in quality trimming. In cases where DNA fragments are cloned before sequencing, the resulting sequence may contain parts of the cloning vector. In contrast, PCR-based cloning and emerging sequencing technologies based on pyrosequencing often avoid using cloning vectors. Recently, one-step Sanger sequencing (combined amplification and sequencing) methods such as Ampliseq and SeqSharp have been developed that allow rapid sequencing of target genes without cloning or prior amplification. Current methods can directly sequence only relatively short (300–1000 nucleotides long) DNA fragments in a single reaction. The main obstacle to sequencing DNA fragments above this size limit is insufficient power of separation for resolving large DNA fragments that differ in length by only one nucleotide. In all cases the use of a primer with a free 3' end is essential. Automated DNA-sequencing instruments (DNA sequencers) can sequence up to 384 DNA samples in a single batch (run) in up to 24 runs a day. DNA sequencers carry out capillary electrophoresis for size separation, detection and recording of dye fluorescence, and data output as fluorescent peak trace chromatograms. Sequencing reactions by thermocycling, cleanup and re-suspension in a buffer solution before loading onto the sequencer are performed separately. A number of commercial and non-commercial software packages can trim low-quality DNA traces automatically. These programs score the quality of each peak and remove low-quality base peaks (generally located at the ends of the sequence). The accuracy of such algorithms is below visual examination by a human operator, but sufficient for automated processing of large sequence data sets.

Amplification and clonal selection

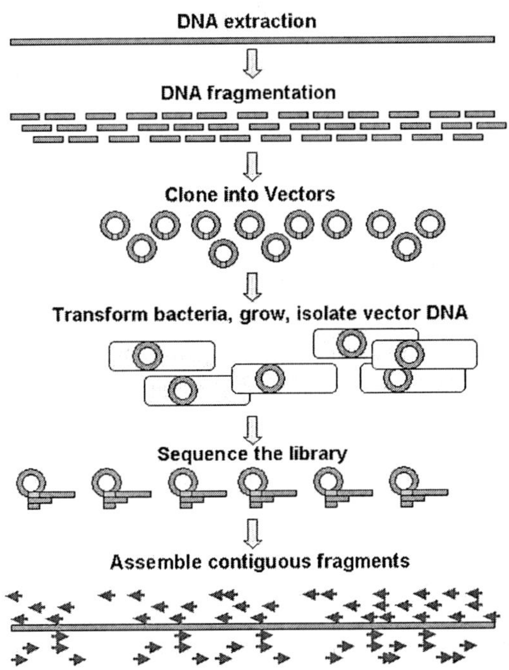

Figure: Genomic DNA is fragmented into random pieces and cloned as a bacterial library. DNA from individual bacterial clones is sequenced and the sequence is assembled by using overlapping DNA regions.

Large-scale sequencing often aims at sequencing very long DNA pieces, such as whole chromosomes, although large-scale sequencing can also be used to generate very large numbers of short sequences, such as found in phage display. For longer targets, such as chromosomes, common approaches consist of cutting (with restriction enzymes) or shearing (with mechanical forces) large DNA fragments into shorter DNA fragments. The fragmented DNA is cloned into a DNA vector, and amplified in Escherichia coli. Short DNA fragments purified from individual bacterial colonies are individually sequenced and assembled electronically into one long, contiguous sequence.

This method does not require any pre-existing information about the sequence of the DNA and is referred to as de novo sequencing. Gaps in the assembled sequence may be filled by primer walking. The different strategies have different tradeoffs in speed and accuracy; shotgun methods are often used for sequencing large genomes, but its assembly is complex and difficult, particularly with sequence repeats often causing gaps in genome assembly.

Most sequencing approaches use an in vitro cloning step to amplify individual DNA molecules, because their molecular detection methods are not sensitive enough for single molecule sequencing. Emulsion PCR isolates individual DNA molecules along with primer-coated beads in aqueous droplets within an oil phase. Polymerase chain reaction (PCR) then coats each bead with clonal copies of the DNA molecule followed by immobilization for later sequencing. Emulsion PCR is used in the methods by Marguilis et al. (commercialized by 454 Life Sciences), Shendure and Porreca et al. (also known as "Polony sequencing") and SOLiD sequencing, (developed by Agencourt, now Applied Biosystems).

Another method for in vitro clonal amplification is bridge PCR, where fragments are amplified upon primers attached to a solid surface, used in the Illumina Genome Analyzer. Single-molecule methods, such as that developed by Stephen Quake's laboratory (later commercialized by Helicos) is an exception: it uses bright fluorophores and laser excitation to detect base addition events from individual DNA molecules fixed to a surface, eliminating the need for molecular amplification.

High-throughput Sequencing

The high demand for low-cost sequencing has driven the development of high-throughput sequencing technologies that parallelize the sequencing process, producing thousands or millions of sequences at once. High-throughput sequencing technologies are intended to lower the cost of DNA sequencing beyond what is possible with standard dye-terminator methods. Massively parallel signature sequencing

Massive parallel signature sequencing (MPSS) is a sequenced based approach that can be used to identify and quantify mRNA transcripts present in a sample similar to serial analysis of gene expression (SAGE) but the biochemical manipulation and sequencing approach differ substantially.

MPSS allows mRNA transcripts to be identified through the generation of a 17-20 bp (base pair) signature sequence adjacent to the 3'-end of the 3'- most site of the designated restriction enzyme (commonly Sau3A or DpnII). Each signature sequence is cloned onto one of a million microbeads. The technique ensures that only one type of DNA sequence is on a microbead. So if there are 50 copies of a specific transcript in the biological sample, these transcripts will be captured onto 50 different microbeads, each bead holding roughly 100 000 amplified copies of the specific signature. The microbeads are then arrayed in a flow cell for sequencing and quantification. The sequence signatures are deciphered by the parallel identification of four bases by hybridization to fluorescently labeled encoders (Figure 5). Each of the encoders has a unique label which is detected after hybridization by taking an image of the microbead array. The next step is to cleave and remove that set of four bases and reveal the next four bases for a new round of hybridization to encoders and image acquisition. The raw output is a list of 17-20 bp signature sequences, that can be annotated to the human genome for gene identification. The longer tag sequence confers a higher specificity than the classical SAGE tag of 9-10 bp. The level of unique gene expression is represented by the count of transcripts present per million molecules, similar to SAGE output. A significant advantage is the larger library size compared with SAGE. An MPSS library typically holds 1 million signature tags, which is roughly 20 times the size of a SAGE library. Some of the disadvantages related to SAGE apply to MPSS as well, such as loss of certain transcripts due to lack of restriction enzyme recognition site and ambiguity in tag annotation. The high sensitivity and absolute gene expression certainly favors MPSS. However, the technology is only available through Lynxgen Therapeutics, Inc. (then Solexa Inc till 2006 and then Illumina). Massively Parallel Signature Sequencing (MPSS) Introduction Massively Parallel Signature Sequencing (MPSS) is an open-ended platform that analyses the level of gene expression in a sample by counting the number of individual mRNA molecules produced by each gene. Tagged PCR products produced from cDNA are amplified so that each corresponding mRNA molecule gives ~100,000 of PCR products with a unique tag. Tags are used to attach the PCR products to microbeads. After several rounds of ligation-based sequence determination using the type IIs restriction endonuclease BbvI, a sequence signature of 16-20 bp is identified from each bead; routinely, 17 bp of high quality sequence is obtained. This is performed in parallel, and approximately 1,000,000 sequence signatures are obtained per experiment. Each signature sequence (MPSS tag) in a MPSS dataset is analyzed, compared with all other signatures and all identical signatures are counted. The level of expression of any single gene is calculated by dividing the number of signatures from that gene by the total number of signatures for all mRNAs present in the dataset. MPSS datasets are additive in nature, which means that datasets from multiple analyses with the same starting mRNA sample can be combined.

MPSS has routine sensitivity at a level of a few molecules of mRNA per cell, and the datasets are in a digital format that simplifies the management and analysis of the data.

How it Works

cDNA fragments are cloned onto micro-beads using the Lynx Megaclone technology. Starting with ome million mRNA molecules from a particular cell or tissue sample, Megaclone will produce one million beads, each containing 100,000 cloned copies of cDNA from each mRNA molecule. All molecules are covalently attached to the micro-beads at their poly(A) ends, so the DpnII end is available for the sequencing reactions.

Lynx Therapeutics' Massively Parallel Signature Sequencing (MPSS)

The first of the "next-generation" sequencing technologies, MPSS was developed in the 1990s at Lynx Therapeutics, a company founded in 1992 by Sydney Brenner and Sam Eletr. MPSS was a bead-based method that used a complex approach of adapter ligation followed by adapter decoding, reading the sequence in increments of four nucleotides; this method made it susceptible to sequence-specific bias or loss of specific sequences. Because the technology was so complex, MPSS was only performed 'in-house' by Lynx Therapeutics and no machines were sold; when the merger with Solexa later led to the development of sequencing-by-synthesis, a more simple approach with numerous advantages, MPSS became obsolete. However, the essential properties of the MPSS output were typical of later "next-gen" data types, including hundreds of thousands of short DNA sequences. In the case of MPSS, these were typically used for sequencing cDNA for measurements of gene expression levels. Lynx Therapeutics merged with Solexa in 2004, and this company was later purchased by Illumina.

Polony Sequencing

Polony sequencing, developed in the laboratory of George Church at Harvard, was among the first next-generation sequencing systems used to sequence a full genome in 2005. It combined an in vitro paired-tag library with emulsion PCR, an automated microscope, and ligation-based sequencing chemistry to sequence an E. coli genome at an accuracy of > 99.9999% and a cost approximately 1/10 that of Sanger sequencing. The technology was licensed to Agencourt Biosciences, subsequently spun out into Agencourt Personal Genomics, and ultimately incorporated into the Applied Biosystems SOLiD platform.

454 Pyrosequencing

454 Life Sciences, is a biotechnology company based in Branford, Connecticut. It is a subsidiary of Roche, and specializes in high-throughput DNA sequencing.

A parallelized version of pyrosequencing was developed by 454 Life Sciences, which has since been acquired by Roche Diagnostics. The method amplifies DNA inside water droplets in an oil solution (emulsion PCR), with each droplet containing a single DNA template attached to a single primer-coated bead that then forms a clonal colony. The sequencing machine contains many picolitre-volume wells each containing a single bead and sequencing enzymes.

Pyrosequencing uses luciferase to generate light for detection of the individual nucleotides added to the nascent DNA, and the combined data are used to generate sequence read-outs. This technology provides intermediate read length and price per base compared to Sanger sequencing on one end and Solexa and SOLiD on the other. History and Major Achievements

454 Life Sciences was founded by Jonathan Rothberg originally as 454 Corporation, a subsidiary of CuraGen Corporation. For their method for low-cost gene sequencing, 454 Life Sciences were awarded the Wall Street Journal's Gold Medal for Innovation in the Biotech-Medical category in 2005. The name 454 was the code name by which the project was referred to at CuraGen and the numbers have no special meaning. In late March, 2007, Roche Diagnostics announced an agreement to purchase 454 Life Sciences for US$154.9 million. It will remain a separate business unit.

In November 2006, Rothberg, Michael Egholm, and colleagues at 454 published a cover article with Svante Paabo in Nature describing the first million base pairs of the Neanderthal genome, and initiated the Neanderthal Genome Project to complete the sequence of the Neanderthal genome by 2009.

In May 2007, Project "Jim", a project initiated by Rothberg and 454 Life Sciences to determine the first sequence of an individual was completed after sequencing the genome of James Dewey Watson, co-discoverer of the structure of DNA.

Technology

454 Sequencing uses a large-scale parallel pyrosequencing system capable of sequencing roughly 400-600 megabases of DNA per 10-hour run on the Genome Sequencer FLX with GS FLX Titanium series reagents. The system relies on fixing nebulized and adapter-ligated DNA fragments to small DNA-capture beads in a water-in-oil emulsion. The DNA fixed to these beads is then amplified by PCR. Each DNA-bound bead is placed into a ~29 ìm well on a PicoTiterPlate, a fiber optic chip. A mix of enzymes such as DNA polymerase, ATP sulfurylase, and luciferase are also packed into the well. The PicoTiterPlate is then placed into the GS FLX System for sequencing.

454 has experienced rapid growth since its acquisition by Roche Diagnostics and release of the GS20 sequencing machine in 2005, the first next-generation DNA sequencer on the market. In 2008, 454 Sequencing launched the GS FLX Titanium series reagents for use on the Genome Sequencer FLX instrument, with the ability to sequence 400-600 million base pairs per run with 400-500 base pair read lengths. The company has said it plans to launch kits enabling sequencing read lengths of up to 1,000 bp in 2010. In late 2009, 454 Life Sciences introduced the GS Junior System, a bench top version of the Genome Sequencer FLX System.

DNA Library Preparation and emPCR

Genomic DNA is fractionated into smaller fragments (300-800 base pairs) and polished (made blunt at each end). Short adaptors are then ligated onto the ends of the fragments. These adaptors provide priming sequences for both amplification and sequencing of the sample-library fragments. One adaptor (Adaptor B) contains a 5'-biotin tag for immobilization of the DNA library onto streptavidin-coated beads. After nick repair, the non-biotinylated strand is released and used as a single-stranded template DNA (sstDNA) library. The sstDNA library is assessed for its quality and the optimal amount (DNA copies per bead) needed for emPCR is determined by titration. The sstDNA library is immobilized onto beads. The beads containing a library fragment carry a single sstDNA molecule. The bead-bound library is emulsified with the amplification reagents in a water-in-oil mixture. Each bead is captured within its own microreactor where PCR amplification occurs. This results in bead-immobilized, clonally amplified DNA fragments.

Sequencing

Single-stranded template DNA library beads are added to the DNA Bead Incubation Mix (containing DNA polymerase) and are layered with Enzyme Beads (containing sulfurylase and luciferase) onto a PicoTiterPlate device. The device is centrifuged to deposit the beads into the wells. The layer of Enzyme Beads ensures that the DNA beads remain positioned in the wells during the sequencing reaction. The bead-deposition process is designed to maximize the number of wells that contain a single amplified library bead.

The loaded PicoTiterPlate device is placed into the Genome Sequencer FLX Instrument. The fluidics sub-system delivers sequencing reagents (containing buffers and nucleotides) across the wells of the plate. The four DNA nucleotides are added sequentially in a fixed order across the PicoTiterPlate device during a sequencing run. During the nucleotide flow, millions of copies of DNA bound to each of the beads are sequenced in parallel.

When a nucleotide complementary to the template strand is added into a well, the polymerase extends the existing DNA strand by adding nucleotide(s). Addition of one (or more) nucleotide(s) generates a light signal that is recorded by the CCD camera in the instrument. This technique is based on sequencing-by-synthesis and is called pyrosequencing. The signal strength is proportional to the number of nucleotides; for example, homopolymer stretches, incorporated in a single nucleotide flow generate a greater signal than single nucleotides. However, the signal strength for homopolymer stretches is linear only up to eight consecutive nucleotides after which the signal falls-off rapidly. Data are stored in standard flowgram format (SFF) files for downstream analysis.

Applications

454 Sequencing can sequence any double-stranded DNA and enables a variety of applications including de novo whole genome sequencing, re-sequencing of whole genomes and target DNA regions, metagenomics and RNA analysis.

Full Genome Sequencing (de novo sequencing and resequencing)

Full genome sequencing (FGS), also referred to as whole genome sequencing (WGS), aims to sequence the entire genome of an organism, for example, humans, dogs, mice, viruses or bacteria. In June 2006, 454 Life Sciences launched a project with the Max Planck Institute for Evolutionary Anthropology to sequence the genome of the Neanderthal, the extinct closest relative of humans.. In September 2008 the complete Neanderthal mitochondrial genome was sequenced, establishing the divergence between humans and Neanderthal at 660,000 +/- 140,000 years, and the full genome was published in 2010, using a combination of 454 and Illumina sequencing.

Amplicon Sequencing

Amplicon (ultra deep) sequencing is a new field which is largely being enabled through 454 Sequencing technology. This method is designed to allow mutations to be detected at extremely low levels, and PCR amplify specific, targeted regions of DNA. This method is used to identify low frequency somatic mutations in cancer samples or discovery of rare variants in HIV infected individuals.

Transcriptome Sequencing

Transcriptome sequencing encompasses experiments including small RNA profiling and discovery, mRNA transcript expression analysis (full-length mRNA, expressed sequence tags (ESTs) and ditags, and allele-specific expression) and the sequencing and analysis of full-length mRNA transcripts. The transcriptome data derived from the Genome Sequencer FLX is ideally suited to detailed transcriptome investigation in the areas of novel gene discovery, gene space identification in novel genomes, assembly of full-length genes, single nucleotide polymorphism (SNP), insertion-deletion and splice-variant discovery.

Metagenomics

Metagenomics is the study of the genomic content in a complex sample. The two primary goals of this approach are to characterize the organisms present in a sample and identify what roles each organism has within a specific environment. Metagenomics samples are found nearly everywhere, including several microenvironments within the human body, soil samples, extreme environments such as deep mines and the various layers within the ocean. The Genome Sequencer FLX System enables a comprehensive view into the diversity of an environmental habitat. The system's long reads ensure the enormous specificity needed to compare sequenced reads against DNA or protein databases. The platform is used to count environmental gene tags to analyze the relative abundance of microbial species under varying environmental conditions.

Illumina (Solexa) Sequencing

Solexa, now part of Illumina, developed a sequencing technology based on reversible dye-terminators. DNA molecules are first attached to primers on a slide and amplified so that local clonal colonies are formed (bridge amplification).

Four types of ddNTPs are added, and non-incorporated nucleotides are washed away. Unlike pyrosequencing, the DNA can only be extended one nucleotide at a time. A camera takes images of the fluorescently labeled nucleotides, then the dye along with the terminal 3' blocker is chemically removed from the DNA, allowing the next cycle.

SOLiD Sequencing

Applied Biosystems' SOLiD technology employs sequencing by ligation. Here, a pool of all possible oligonucleotides of a fixed length are labeled according to the sequenced position. Oligonucleotides are annealed and ligated; the preferential ligation by DNA ligase for matching sequences results in a signal informative of the nucleotide at that position. Before sequencing, the DNA is amplified by emulsion PCR. The resulting bead, each containing only copies of the same DNA molecule, are deposited on a glass slide. The result is sequences of quantities and lengths comparable to Illumina sequencing. ABI Solid Sequencing

SOLiD (Sequencing by Oligonucleotide Ligation and Detection) is a next-generation sequencing technology developed by Life Technologies and has been commercially available since 2008. These next generation technologies generate hundreds of millions to billions of small sequence reads at one time.

Well-known examples of such DNA sequencing methods include 454 pyrosequencing (introduced in 2005, generating millions of 200-400bp reads in 2009), the Solexa system (introduced in 2006, generating hundreds of millions of 50-100bp reads in 2009) and the SOLiD system (introduced in 2007, generating billions of 50bp reads in 2009). These methods have reduced the cost from $0.01/base in 2004 to nearly $0.0001/base in 2006 and increased the sequencing capacity from 1,000,000 bases/machine/day in 2004 to more than 5,000,000,000 bases/machine/day in 2009. Over 30 publications exist describing its use first for nucleosome positioning from Valouev et al., transcriptional profiling or strand sensitive RNA-Seq with Cloonan et al., single cell transcriptional profiling with Tang et al. and ultimately human resequencing with McKernan et al.

Chemistry

A library of DNA fragments is prepared from the sample to be sequenced, and are used to prepare clonal bead populations. That is, only one species of fragment will be present on the surface of each magnetic bead. The fragments attached to the magnetic beads will have a universal P1 adapter sequence attached so that the starting sequence of every fragment is both known and identical. Emulsion PCR takes place in microreactors containing all the necessary reagents for PCR. The resulting PCR products attached to the beads are then covalently bound to a glass slide.

Primers hybridize to the P1 adapter sequence within the library template. A set of four fluorescently labeled di-base probes compete for ligation to the sequencing primer. Specificity of the di-base probe is achieved by interrogating every 1st and 2nd base in each ligation reaction. Multiple cycles of ligation, detection and cleavage are performed with the number of cycles determining the eventual read length. Following a series of ligation cycles, the extension product is removed and the template is reset with a primer complementary to the n-1 position for a second round of ligation cycles.

Five rounds of primer reset are completed for each sequence tag. Through the primer reset process, each base is interrogated in two independent ligation reactions by two different primers. For example, the base at read position 5 is assayed by primer number 2 in ligation cycle 2 and by primer number 3 in ligation cycle 1.

4

Methods Based on Knowledge

Regulated Stages of Gene Expression

Any step of gene expression may be modulated, from the DNA-RNA transcription step to post-translational modification of a protein. The following is a list of stages where gene expression is regulated, the most extensively utilised point is Transcription Initiation:

- Chromatin domains
- Transcription
- Post-transcriptional modification
- RNA transport
- Translation
- mRNA degradation

Modification of DNA

In eukaryotes, the accessibility of large regions of DNA can depend on its chromatin structure, which can be altered as a result of histone modifications directed by DNA methylation, ncRNA, or DNA-binding protein. Hence these modifications may up or down regulate the expression of gene. Certain of these modifications that regulate gene expression are inheritable and are referred to as epigenetic regulation.

Chemical

Methylation of DNA is a common method of gene silencing. DNA is typically methylated by methyltransferase enzymes on cytosine nucleotides in a CpG dinucleotide sequence (also called "CpG islands" when densely clustered). Analysis of the pattern of methylation in a given region of DNA (which can be a promoter) can be achieved through a method called bisulfite mapping. Methylated cytosine residues are unchanged by the treatment, whereas unmethylated ones are changed to uracil. The differences are analyzed by DNA sequencing or by methods developed to quantify SNPs, such as Pyrosequencing (Biotage) or MassArray (Sequenom), measuring the relative amounts of C/T at the CG dinucleotide. Abnormal methylation patterns are thought to be involved in oncogenesis.

Structural

Transcription of DNA is dictated by its structure. In general, the density of its packing is indicative of the frequency of transcription. Octameric protein complexes called nucleosomes are responsible for the amount of supercoiling of DNA, and these complexes can be temporarily modified by processes such as phosphorylation or more permanently modified by processes such as methylation. Such modifications are considered to be responsible for more or less permanent changes in gene expression levels.

Histone acetylation is also an important process in transcription. Histone acetyltransferase enzymes (HATs) such as CREB-binding protein also dissociate the DNA from the histone complex, allowing transcription to proceed. Often, DNA methylation and histone deacetylation work together in gene silencing. The combination of the two seems to be a signal for DNA to be packed more densely, lowering gene expression.

Regulation of Transcription

Regulation of transcription controls when transcription occurs and how much RNA is created. Transcription of a gene by RNA polymerase can be regulated by at least five mechanisms:

- Specificity factors alter the specificity of RNA polymerase for a given promoter or set of promoters, making it more or less likely to bind to them (i.e., sigma factors used in prokaryotic transcription).
- Repressors bind to non-coding sequences on the DNA strand that are close to or overlapping the promoter region, impeding RNA polymerase's progress along the strand, thus impeding the expression of the gene.
- General transcription factors position RNA polymerase at the start of a protein-coding sequence and then release the polymerase to transcribe the mRNA.
- Activators enhance the interaction between RNA polymerase and a particular promoter, encouraging the expression of the gene. Activators do this by increasing the attraction of RNA

polymerase for the promoter, through interactions with subunits of the RNA polymerase or indirectly by changing the structure of the DNA.

- Enhancers are sites on the DNA helix that are bound to by activators in order to loop the DNA bringing a specific promoter to the initiation complex. Enhancers are much more common in eukaryote than prokaryotes, where only a few examples exist (to date).

Post-transcriptional Regulation

After the DNA is transcribed and mRNA is formed, there must be some sort of regulation on how much the mRNA is translated into proteins. Cells do this by modulating the capping, splicing, addition of a Poly(A) Tail, the sequence-specific nuclear export rates, and, in several contexts, sequestration of the RNA transcript. These processes occur in eukaryotes but not in prokaryotes. This modulation is a result of a protein or transcript that, in turn, is regulated and may have an affinity for certain sequences.

Regulation of Translation

The translation of mRNA can also be controlled by a number of mechanisms, mostly at the level of initiation. Recruitment of the small ribosomal subunit can indeed be modulated by mRNA secondary structure, antisense RNA binding, or protein binding. In both prokaryotes and eukaryotes, a large number of RNA binding proteins exist, which often are directed to their target sequence by the secondary structure of the transcript, which may change depending on certain conditions, such as temperature or presence of a ligand (aptamer). Some transcripts act as ribozymes and self-regulate their expression.

Examples of Gene Regulation

- Enzyme induction is a process in which a molecule (e.g., a drug) induces (i.e., initiates or enhances) the expression of an enzyme.
- The induction of heat shock proteins in the fruit fly Drosophila melanogaster.
- The Lac operon is an interesting example of how gene expression can be regulated.
- Viruses, despite having only a few genes, possess mechanisms to regulate their gene expression, typically into an early and late phase, using collinear systems regulated by anti-terminators (lambda phage) or splicing modulators (HIV).

Developmental Biology

A large number of studied regulatory systems come from developmental biology. Examples include:

- The colinearity of the Hox gene cluster with their nested antero-posterior patterning
- It has been speculated that pattern generation of the hand (digits - interdigits) The gradient of Sonic hedgehog (secreted inducing factor) from the zone of polarizing activity in the limb, which creates a gradient of active Gli3, which activates Gremlin, which inhibits BMPs also secreted in the limb, resulting in the formation of an alternating pattern of activity as a result of this reaction-diffusion system.
- Somitogenesis is the creation of segments (somites) from a uniform tissue (Pre-somitic Mesoderm, PSM). They are formed sequentially from anterior to posterior. This is achieved in amniotes possibly by means of two opposing gradients, Retinoic acid in the anterior (wavefront) and Wnt and Fgf in the posterior, coupled to an oscillating pattern (segmentation clock) composed of FGF + Notch and Wnt in antiphase.
- Sex determination in the soma of a Drosophila requires the sensing of the ratio of autosomal genes to sex chromosome-encoded genes, which results in the production of sexless splicing factor in females, resulting in the female isoform of doublesex.

Morphogen

A morphogen is a substance governing the pattern of tissue development, and the positions of the various specialized cell types within a tissue. It spreads from a localized source and forms a concentration gradient across a developing tissue .

In developmental biology a morphogen is rigorously used to mean a signaling molecule that acts directly on cells (not through serial induction) to produce specific cellular responses dependent on morphogen concentration .

Well-known morphogens include: Decapentaplegic / Transforming growth factor beta, Hedgehog / Sonic Hedgehog, Wingless / Wnt, Epidermal growth factor, Fibroblast growth factor, and Retinoic acid.

Morphogens are defined conceptually, not chemically, so simple compounds such as retinoic acid (the active metabolite of retinol or vitamin A) may also act as morphogens

Introduction

During early development, morphogen gradients generate different cell types in distinct spatial order. The morphogen provides spatial information by forming a concentration gradient that subdivides a field of cells by inducing or maintaining the expression of different target genes at distinct concentration thresholds. Thus, cells far from the source of the morphogen will receive low levels of morphogen and express only low-threshold target genes. In contrast, cells close to the source of morphogen will receive high levels of morphogen and will express both low- and high-threshold target genes. Distinct cell types emerge as a consequence of the different combination of target gene expression. In this way, the field of cells is subdivided into different types according to their position relative to the source of the morphogen. This is a general mechanism by which cell type diversity can be generated in animal development.

Some of the earliest and best-studied morphogens are transcription factors that diffuse within early Drosophila melanogaster (fruit fly) embryos. However, most morphogens are secreted proteins that signal between cells.

Fruit Fly

Drosophila melanogaster has an unusual developmental system, in which the first thirteen cell divisions of the embryo occur within a syncytium prior to cellularization. Essentially the embryo remains a single cell with over 8000 nuclei evenly spaced near the membrane until the fourteenth cell division, when independent membranes furrow between the nuclei, separating them into independent cells. As a result, in fly embryos transcription factors such as Bicoid or Hunchback can act as morphogens because they can freely diffuse between nuclei to produce smooth gradients of concentration without relying on specialized intercellular signalling mechanisms. Although there is some evidence that homeobox transcription factors similar to these can pass directly through cell membranes, this mechanism is not believed to contribute greatly to morphogenesis in cellularized systems.

In most developmental systems, such as human embryos or later Drosophila development, syncytia occur only rarely (such as in skeletal muscle), and morphogens are generally secreted signalling proteins. These proteins bind to the extracellular domains of transmembrane receptor proteins, which use an elaborate process of signal transduction to communicate the level of morphogen to the nucleus. The nuclear targets of signal transduction pathways are usually transcription factors, whose activity is regulated in a manner that reflects the level of morphogen received at the cell surface. Thus, secreted morphogens act to generate gradients of transcription factor activity just like those that are generated in the syncitial Drosophila embryo.

Discrete target genes respond to different thresholds of morphogen activity. The expression of target genes is controlled by segments of DNA called 'enhancers' to which transcription factors bind directly. Once bound, the transcription factor then stimulates or inhibits the transcription of the gene and thus controls the level of expression of the gene product (usually a protein). 'Low-threshold' target genes require only low levels of morphogen activity to be regulated and feature enhancers that contain many high-affinity binding sites for the transcription factor. 'High-threshold' target genes have relatively fewer binding sites or low-affinity binding sites that require much greater levels of transcription factor activity to be regulated.

Thus, the general mechanism by which morphogens subdivide tissues into patterns of distinct cell types is well understood. However, morphogens often have additional activities such as controlling the growth of the tissue or orienting the polarity of cells within it (for example, the hairs on your forearm point in one direction). These activities of morphogens are much less understood and are the subject of current research efforts in the field of developmental biology.

Discovery

The morphogen idea has a long history in developmental biology, dating back to the work of the pioneering Drosophila geneticist, Thomas Hunt Morgan, in the early 20th century. However, it was Lewis Wolpert who refined the morphogen concept in the 1960s with his famous French flag model which described how morphogen could subdivide a tissue into domains of different target gene expression (corresponding to the colours of the French flag).

This model was championed by the leading Drosophila biologist, Peter Lawrence. Christiane Nusslein-Volhard identified the first morphogen, Bicoid, one of the transcription factors present in a gradient in the Drosophila syncitial embryo. Two labs, that of Gary Struhl and that of Stephen Cohen, then demonstrated that a secreted signalling protein, Decapentaplegic (the Drosophila homologue of Transforming Growth Factor Beta), acted as a morphogen during later stages of Drosophila development. Subsequent studies of the development of many different animals has confirmed the widespread importance of morphogens in governing animal development.

Chapter. Gene Regulatory Network

A gene regulatory network or genetic regulatory network (GRN) is a collection of DNA segments in a cell which interact with each other indirectly (through their RNA and protein expression products) and with other substances in the cell, thereby governing the rates at which genes in the network are transcribed into mRNA. In general, each mRNA molecule goes on to make a specific protein (or set of proteins). In some cases this protein will be structural, and will accumulate at the cell membrane or within the cell to give it particular structural properties. In other cases the protein will be an enzyme, i.e., a micro-machine that catalyses a certain reaction, such as the breakdown of a food source or toxin. Some proteins though serve only to activate other genes, and these are the transcription factors that are the main players in regulatory networks or cascades. By binding to the promoter region at the start of other genes they turn them on, initiating the production of another protein, and so on. Some transcription factors are inhibitory.

In single-celled organisms, regulatory networks respond to the external environment, optimising the cell at a given time for survival in this environment. Thus a yeast cell, finding itself in a sugar solution, will turn on genes to make enzymes that process the sugar to alcohol. This process, which we associate with wine-making, is how the yeast cell makes its living, gaining energy to multiply, which under normal circumstances would enhance its survival prospects.

In multicellular animals the same principle has been put in the service of gene cascades that control body-shape. Each time a cell divides, two cells result which, although they contain the same genome in full, can differ in which genes are turned on and making proteins. Sometimes a 'self-sustaining feedback loop' ensures that a cell maintains its identity and passes it on. Less understood is the mechanism of epigenetics by which chromatin modification may provide cellular memory by blocking or allowing transcription. A major feature of multicellular animals is the use of morphogen gradients, which in effect provide a positioning system that tells a cell where in the body it is, and hence what sort of cell to become. A gene that is turned on in one cell may make a product that leaves the cell and diffuses through adjacent cells, entering them and turning on genes only when it is present above a certain threshold level. These cells are thus induced into a new fate, and may even generate other morphogens that signal back to the original cell. Over longer distances morphogens may use the active process of signal transduction. Such signalling controls embryogenesis, the building of a body plan from scratch through a series of sequential steps. They also control maintain adult bodies through feedback processes, and the loss of such feedback because of a mutation can be responsible for the cell proliferation that is seen in cancer. In parallel with this process of building structure, the gene cascade turns on genes that make structural proteins that give each cell the physical properties it needs. It has been suggested that, because biological molecular interactions are intrinsically stochastic, gene networks are the result of cellular processes and not their cause (i.e. Cellular Darwinism). However, recent experimental evidence has favored the attractor view of cell fates.

Overview

At one level, biological cells can be thought of as "partially-mixed bags" of biological chemicals – in the discussion of gene regulatory networks, these chemicals are mostly the mRNAs and proteins that arise from gene expression. These mRNA and proteins interact with each other with various degrees of specificity. Some diffuse around the cell. Others are bound to cell membranes, interacting with molecules in the environment.

Still others pass through cell membranes and mediate long range signals to other cells in a multi-cellular organism. These molecules and their interactions comprise a gene regulatory network. A typical gene regulatory network looks something like this:

The nodes of this network are proteins, their corresponding mRNAs, and protein/protein complexes. Nodes that are depicted as lying along vertical lines are associated with the cell/environment interfaces, while the others are free-floating and diffusible. Implied are genes, the DNA sequences which are transcribed into the mRNAs that translate into proteins.

Edges between nodes represent individual molecular reactions, the protein/protein and protein/mRNA interactions through which the products of one gene affect those of another, though the lack of experimentally obtained information often implies that some reactions are not modeled at such a fine level of detail. These interactions can be inductive (the arrowheads), with an increase in the concentration of one leading to an increase in the other, or inhibitory (the filled circles), with an increase in one leading to a decrease in the other. A series of edges indicates a chain of such dependences, with cycles corresponding to feedback loops. The network structure is an abstraction of the system's chemical dynamics, describing the manifold ways in which one substance affects all the others to which it is connected. In practice, such GRNs are inferred from the biological literature on a given system and represent a distillation of the collective knowledge about a set of related biochemical reactions.

Genes can be viewed as nodes in the network, with input being proteins such as transcription factors, and outputs being the level of gene expression. The node itself can also be viewed as a function which can be obtained by combining basic functions upon the inputs (in the Boolean network described below these are Boolean functions, typically AND, OR, and NOT). These functions have been interpreted as performing a kind of information processing within the cell, which determines cellular behavior.

The basic drivers within cells are concentrations of some proteins, which determine both spatial (location within the cell or tissue) and temporal (cell cycle or developmental stage) coordinates of the cell, as a kind of "cellular memory". The gene networks are only beginning to be understood, and it is a next step for biology to attempt to deduce the functions for each gene "node", to help understand the behavior of the system in increasing levels of complexity, from gene to signaling pathway, cell or tissue level.

Mathematical models of GRNs have been developed to capture the behavior of the system being modeled, and in some cases generate predictions corresponding with experimental observations. In some other cases, models have proven to make accurate novel predictions, which can be tested experimentally, thus suggesting new approaches to explore in an experiment that sometimes wouldn't be considered in the design of the protocol of an experimental laboratory. The most common modeling technique involves the use of coupled ordinary differential equations (ODEs). Several other promising modeling techniques have been used, including Boolean networks, Petri nets, Bayesian networks, graphical Gaussian models, Stochastic, and Process Calculi. Conversely, techniques have been proposed for generating models of GRNs that best explain a set of time series observations.

Modelling

Coupled ODEs

For an example of modelling of the cell cycle with ODEs, see cellular model. It is common to model such a network with a set of coupled ordinary differential equations (ODEs) or stochastic ODEs, describing the reaction kinetics of the constituent parts. Suppose that our regulatory network has N nodes, and let $S_1(t), S_2(t), ..., S_N(t)$ represent the concentrations of the N corresponding substances at time t. Then the temporal evolution of the system can be described approximately by

$$\frac{dS_j}{dt} = f_j(S_1, S_2, ..., S_N)$$

Where the functions f_j express the dependence of S_j on the concentrations of other substances present in the cell. The functions f_j are ultimately derived from basic principles of chemical kinetics or simple expressions derived from these e.g. Michaelis-Menten enzymatic kinetics. Hence, the functional forms of the f_j are usually chosen as low-order polynomials or Hill functions that serve as an ansatz for the real molecular dynamics. Such models are then studied using the mathematics of nonlinear dynamics. System-specific information, like reaction rate constants and sensitivities, are encoded as constant parameters.

By solving for the fixed point of the system:

$$\frac{dS_j}{dt} = 0$$

For all j, one obtains (possibly several) concentration profiles of proteins and mRNAs that are theoretically sustainable (though not necessarily stable). Steady states of kinetic equations thus correspond to potential cell types, and oscillatory solutions to the above equation to naturally cyclic cell types. Mathematical stability of these attractors can usually be characterized by the sign of higher derivatives at critical points, and then correspond to biochemical stability of the concentration profile.

Critical points and bifurcations in the equations correspond to critical cell states in which small state or parameter perturbations could switch the system between one of several stable differentiation fates. Trajectories correspond to the unfolding of biological pathways and transients of the equations to short-term biological events. For a more mathematical discussion, see the articles on nonlinearity, dynamical systems, bifurcation theory, and chaos theory.

Boolean Network

The following example illustrates how a Boolean network can model a GRN together with its gene products (the outputs) and the substances from the environment that affect it (the inputs). Stuart Kauffman was amongst the first biologists to use the metaphor of Boolean networks to model genetic regulatory networks.

1. Each gene, each input, and each output is represented by a node in a directed graph in which there is an arrow from one node to another if and only if there is a causal link between the two nodes.
2. Each node in the graph can be in one of two states: on or off.
3. For a gene, "on" corresponds to the gene being expressed; for inputs and outputs, "on" corresponds to the substance being present.
4. Time is viewed as proceeding in discrete steps. At each step, the new state of a node is a Boolean function of the prior states of the nodes with arrows pointing towards it.

The validity of the model can be tested by comparing simulation results with time series observations.

Continuous Networks

Continuous network models of GRNs are an extension of the boolean networks described above. Nodes still represent genes and connections between them regulatory influences on gene expression. Genes in biological systems display a continuous range of activity levels and it has been argued that using a continuous representation captures several properties of gene regulatory networks not present in the Boolean model. Formally most of these approaches are similar to an artificial neural network, as inputs to a node are summed up and the result serves as input to a sigmoid function, e.g., but proteins do often control gene expression in a synergistic, i.e. non-linear, way. However there is now a continuous network model that allows grouping of inputs to a node thus realizing another level of regulation. This model is formally closer to a higher order recurrent neural network. The same model has also been used to mimic the evolution of cellular differentiation and even multicellular morphogenesis.

Stochastic Gene Networks

Recent experimental results have demonstrated that gene expression is a stochastic process. Thus, many authors are now using the stochastic formalism, after the work by. Works on single gene expression and small synthetic genetic networks, such as the genetic toggle switch of Tim Gardner and Jim Collins, provided additional experimental data on the phenotypic variability and the stochastic nature of gene expression. The first versions of stochastic models of gene expression involved only instantaneous reactions and were driven by the Gillespie algorithm.

Since some processes, such as gene transcription, involve many reactions and could not be correctly modeled as an instantaneous reaction in a single step, it was proposed to model these reactions as single step multiple delayed reactions in order to account for the time it takes for the entire process to be complete.

From here, a set of reactions were proposed that allow generating GRNs. These are then simulated using a modified version of the Gillespie algorithm, that can simulate multiple time delayed reactions (chemical reactions where each of the products is provided a time delay that determines when will it be released in the system as a "finished product").

For example, basic transcription of a gene can be represented by the following single-step reaction (RNAP is the RNA polymerase, RBS is the RNA ribosome binding site, and Pro_i is the promoter region of gene i):

$$RNAP + \mathrm{Pr}\,o_i \xrightarrow{k_i,bas} \mathrm{Pr}\,o_i\left(\tau_i^1\right) + RBS_i\left(\tau_i^1\right) + RNAp\left(\tau_I^2\right)$$

A recent work proposed a simulator (SGNSim, Stochastic Gene Networks Simulator), that can model GRNs where transcription and translation are modeled as multiple time delayed events and its dynamics is driven by a stochastic simulation algorithm (SSA) able to deal with multiple time delayed events. The time delays can be drawn from several distributions and the reaction rates from complex functions or from physical parameters. SGNSim can generate ensembles of GRNs within a set of user-defined parameters, such as topology. It can also be used to model specific GRNs and systems of chemical reactions. Genetic perturbations such as gene deletions, gene over-expression, insertions, frame shift mutations can also be modeled as well.

The GRN is created from a graph with the desired topology, imposing in-degree and out-degree distributions. Gene promoter activities are affected by other genes expression products that act as inputs, in the form of monomers or combined into multimers and set as direct or indirect. Next, each direct input is assigned to an operator site and different transcription factors can be allowed, or not, to compete for the same operator site, while indirect inputs are given a target. Finally, a function is assigned to each gene, defining the gene's response to a combination of transcription factors (promoter state). The transfer functions (that is, how genes respond to a combination of inputs) can be assigned to each combination of promoter states as desired.

In other recent work, multiscale models of gene regulatory networks have been developed that focus on synthetic biology applications. Simulations have been used that model all biomolecular interactions in transcription, translation, regulation, and induction of gene regulatory networks, guiding the design of synthetic systems.

Prediction

Other work has focused on predicting the gene expression levels in a gene regulatory network. The approaches used to model gene regulatory networks have been constrained to be interpretable and, as a result, are generally simplified versions of the network. For example, Boolean networks have been used due to their simplicity and ability to handle noisy data but lose data information by having a binary representation of the genes. Also, artificial neural networks omit using a hidden layer so that they can be interpreted, losing the ability to model higher order correlations in the data. Using a model that is not constrained to be interpretable, a more accurate model can be produced. Being able to predict gene expressions more accurately provides a way to explore how drugs affect a system of genes as well as for finding which genes are interrelated in a process. This has been encouraged by the DREAM competition which promotes a competition for the best prediction algorithms. Some other recent work has used artificial neural networks with a hidden layer.

Network Connectivity

Empirical data indicate that biological gene networks are sparsely connected, and that the average number of upstream-regulators per gene is less than two. Theoretical results show that selection for robust gene networks will favor minimally complex, more sparsely connected, networks. These results suggest that a sparse, minimally connected, genetic architecture may be a fundamental design constraint shaping the evolution of gene network complexity.

Body plan

A body plan is the blueprint for the way the body of an organism is laid out. An organism's symmetry, its number of body segments and number of limbs are all aspects of its body plan. One of the key issues of developmental biology is the evolution of body plans as different as those of a starfish, a fern, or a mammal, from a common biological heritage, and in particular how radical changes in body plans have occurred over geological time. The body plan is a key feature of an organism's morphology and, since the discovery of DNA, developmental biologists have been able to learn a lot about how genes control the development of structural features through a cascade of processes in which key genes produce morphogens, chemicals that diffuse through the body to produce a gradient that acts as a position indicator for cells, turning on other genes, some of which in turn produce other morphogens. A key discovery was the existence of groups of homeobox genes which are responsible for laying down the basic body plan in organisms.

The homeobox genes are remarkably conserved between species as diverse as the fruitfly and man, the basic segmented pattern of the worm or fruitfly being the origin of the segmented spine in man. The field of evolutionary developmental biology, which studies the genetics of morphology in detail is now a rapidly expanding one , with many of the developmental genetic cascades, particularly in the fruitfly drosophila, now catalogued in considerable detail.

Body plan is the basis for phylum, and there are 35 different basic animal body plans, corresponding to different phyla.

Origin

The evolution of body plans became inevitable with the emergence of differentiated multicellular life in the Ediacaran Era, over 600 million years ago. The most basic and successful structure, for free-moving organisms, is the "pipe" or alimentary canal. This is common even to organisms as diverse as humans and earthworms.

It is essentially a passage having a mouth at one end, and a cloaca or anus at the other. The simple process of nutrient capture, digestion, and waste disposal is fundamental to the body plan of advanced, free-moving animals. Vertebra, limbs, even brains are supplementary to the pipe. Natural selection has spun off an enormous range of variations on this basic theme, but the pipe model itself remains. The basic symmetry and organization of this body plan apparently gave an ancient organism an enormous advantage at survival and reproduction, and it has been preserved in most animals ever since.

The Cambrian explosion refers to the massive increase in different body plans that took place around 530 million years ago. Fossils from this era show all sorts of odd shapes, many quite unlike anything found today. At that time it was possible for organisms to survive and make a living even though they were unrefined and unlikely, because predation had yet to evolve, along with arms races that would optimise and streamline them to occupy a particular ecological niche.

Bauplan

Bauplan (German for building plan, blueprint; plural: baupläne or bauplaene) is a closely related term in biology referring to the common new and original (homologous) properties of the members of a systematic group (taxon). It is not necessary that a bauplan precisely describes any one particular species of that group.

The concept of bauplan is employed in the studies of morphology, taxonomy, comparative physiology, evolutionary physiology, and, most notably, phylogenetics and evolution. Before the advent of genetic sequencing, the analysis of the bauplan of fossils was the primary method to determine hypothetical relationships and lineages of species, both living and extinct. The idea is that species that are closely related share more common properties, hence a more detailed bauplan. Small differences of bauplan are indicative of species belonging to a parent, child or sibling taxon.

Genetic Basis

Similarities and differences in adult shape and form, as well as the developmental pattern of embryos, provide the framework for modern taxonomic classification. These comparisons are the basis of phylogenetic systematics. Embryonic development is relatively consistent among animals with similar body plans, although similar larval forms may give rise to very different adults in some groups. The timing, pattern, and scale of developmental events determine the shape of an organism, and closely related groups are more likely to share structural and developmental similarities than those that are more distantly related. Homologous structures and developmental stages—those that are similar among related groups because they are inherited from a common ancestor—are the basis of modern biological lassification.xamples

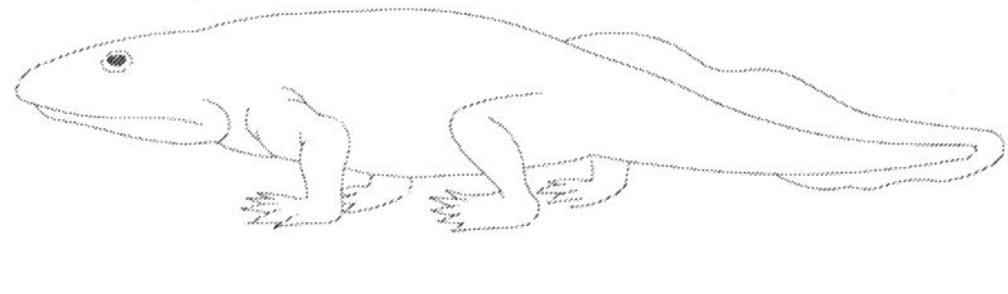

Figure: An early tetrapod: the ichthyostega

The current range of body plans is far from exhausting the possible patterns for life: the Ediacaran biota appears to contain numerous species and taxa with body plans quite different from any found in currently living organisms.

The most commonly seen body plan amongst vertebrates is that of the tetrapod, which include all mammals, birds, amphibians and reptiles. Some animal groups, such as the cetaceans, bats and most birds have been modified (e.g. front limbs become wings or flippers) but nevertheless, they are still tetrapods.

The invertebrates employ a much more diverse array of body plans, such as seen in insects (six legs, three body parts and an exoskeleton), cephalopods (no skeleton, hydrostatically stiffened tentacles, primary propulsion by squeezing water out of a mantle cavity), echinoderms (fivefold radial symmetry, external skeleton, movement by hydrostatically operated tube feet) and various phyla of "worms" (tube-shaped, movement by expanding and contracting parts of the body).

The most varied collection of body forms known is found in the Burgess Shale, where fossils from a Cambrian sea show a tremendous variety of body forms that came to rise (only to later fall extinct) during the Cambrian explosion.

Fictional

One common theme in science fiction is the appearance of extraterrestrial beings, descriptions of which have ranged from being simple variants on human anatomy to beings with body plans wildly different from any found on Earth. The field of exobiology attempts to bring these and similar speculations into the realm of serious scientific investigation.

Cis-regulatory module

Cis-regulatory module (CRM) is a stretch of DNA, usually 100-1000 DNA base pairs in length, where a number of transcription factors can bind and regulate expression of nearby genes. One cis-regulatory element can regulate several genes, and conversely, one gene can have several cis-regulatory modules. Cis-regulatory modules are one of several types of functional regulatory elements. Regulatory elements are binding sites for transcription factors, which are involved in gene regulation. Cis-regulatory modules perform a large amount of developmental information processing. Cis-regulatory modules are non-random clusters at their specified target site that contain transcription factor binding sites.

They are labeled as cis because they are typically located on the same DNA as the genes they control as opposed to trans, which refers to effects on genes not located on the same strand or farther away, such as transcription factors. The original definition presented cis-regulatory modules as enhancers of cis-acting DNA, which increased the rate of transcription from a linked promoter. However, this definition has changed to define cis-regulatory modules as a DNA sequence with transcription factor binding sites which are clustered into modular structures, including -but not limited to- locus control regions, promoters, enhancers, silencers, boundary control elements and other modulators. Cis-regulatory modules can be divided into classes. Enhancers regulate gene expression positively. Insulators work indirectly by interacting with other nearby cis-regulatory modules. Silencers act by silencing genes. Cis-regulatory modules carry out their function by integrating the active transcription factors and the associated co-factors at a specific time and place in the cell, this information is read and an output is given.

Gene-regulation function of a cis-regulatory module

The design of cis-regulatory modules is such that transcription factors and epigenetic modifications serve as inputs, and the output of the module is the command given to the transcription machinery, which in turn determines the rate of gene transcription or whether it is turned on or off. There are two types of transcription factor inputs: those that determine when the target gene is to be expressed and those that serve as functional drivers, which come into play only during specific situations during development. These inputs can come from different time points, can represent different signal ligands, or can come from different domains or lineages of cells. However, a lot still remains unknown. Additionally, the regulation of chromatin structure and nuclear organization also play a role in determining and controlling the function of cis-regulatory modules. Thus gene-regulation functions (GRF) provide a unique characteristic of a cis-regulatory module (CRM), relating the concentrations of transcription factors (input) to the promoter activities (output). The challenge is to predict GRFs. This challenge still remains unsolved. In general, gene-regulation functions do not use Boolean logic, although in some cases the approximation of the Boolean logic is still very useful.

The Boolean Logic Assumption

Within the assumption of the Boolean logic, four principles guiding the operation of these cis-regulatory modules are : 1) The design of the cis-regulatory module determines the regulatory function. 2) In relation to development, these modules can generate both positive and negative outputs. 3) The output of each module is a product of the various operations performed on it. Common operations: "OR" logic gate- This design indicates that in an output will be given when either input is given . "AND" logic gate– In this design two different regulatory factors are necessary to make sure that a positive output results. "Toggle Switches"- This design occurs when the signal ligand is absent while the transcription factor is present; this transcription factor ends up acting as a dominant repressor. However, once the signal ligand is present the transcription factor's role as repressor is eliminated and transcription can occur. Other Boolean logic operations can occur as well, such as sequence specific transcriptional repressors, which when they bind to the cis-regulatory module lead to an output of zero. Additionally, besides influence from the different logic operations, the output of a "cis"-regulatory module will also be influenced by prior events. 4) Cis-regulatory modules must interact with other regulatory elements.

For the most part, even with the presence of functional overlap between cis-regulatory modules of a gene, the modules' inputs and outputs tend to not be the same.

Identification and Computational Prediction of CRMs

Besides experimentally determining CRMs, there are various bioinformatics algorithms for predicting them. Most algorithms try to search for significant combinations of transcription factor binding sites (DNA binding sites) in promoter sequences of co-expressed genes. More advanced methods combine the search for significant motifs with correlation in gene expression datasets between transcription factors and target genes. Both methods have been implemented, for example, in the ModuleMaster. Other programs created for the identification and prediction of "cis-regulatory modules include:

Stubb uses hidden Markov models to identify statistically significant clusters of transcription factor combinations. It also uses a second related genome to improve the prediction accuracy of the model.

Bayesian Networks use an algorithm that combines site predictions and tissue-specific expression data for transcription factors and target genes of interest. This model also uses regression trees to depict the relationship between the identified cis-regulatory module and the possible binding set of transcription factors.

CRÈME examine clusters of target sites for transcription factors of interest. This program uses a database of confirmed transcription factor binding sites that were annotated across the human genome. A search algorithm is applied to the data set to identify possible combinations of transcription factors, which have binding sites that are close to the promoter of the gene set of interest. The possible cis-regulatory modules are then statistically analyzed and the significant combinations are graphically represented

Active cis-regulatory modules in a genomic sequence have been difficult to identify. Problems in identification arise because often scientists find themselves with a small set of known transcription factors, so it makes it harder to identify statistically significant clusters of transcription factor binding sites. Additionally, high costs limit the use of large whole genome tiling arrays.

Classification of Cis-regulatory Modules

Cis-regulatory modules can be characterized by the information processing that they encode and the organization of their transcription factor binding sites. Additionally, cis-regulatory modules are also characterized by the way they affect the probability, proportion, and rate of transcription. Highly cooperative and coordinated cis-regulatory modules are classified as enhanceosomes.

The architecture and the arrangement of the transcription factor binding sites are critical because disruption of the arrangement could cancel out the function. Functional flexible cis-regulatory modules are called billboards. Their transcriptional output is the summation effect of the bound transcription factors. Enhancers affect the probability of a gene being activated, but have little or no effect on rate. Binary response model acts like an on/off switch for transcription. This model will increase or decrease the amount of cells that transcribe a gene, however it does not affect the rate of transcription. Rheostatic response model describes cis-regulatory modules as regulators of the initiation rate of transcription of its associated gene.

Mode of Action

Cis-regulatory modules can regulate their target genes over large distances, several models have been proposed to describe the way that these modules may communicate with their target gene promoter. DNA Scanning Model: In this model the transcription factor and cofactor complex form at the cis-regulatory module and then continues to move along the DNA sequence until it finds the target gene promoter. Looping Model: In this model the transcription factor binds to the cis-regulatory module, which then causes the looping of the DNA sequence and allows for the interaction with the target gene promoter. Facilitated Tracking Model: This model combines parts of the two previous models.

The transcription factor-cis-regulatory module complex causes the looping of the DNA sequence slowly towards the target promoter and forms a stable looped configuration.

Cis-regulatory Module in Gene Regulatory Network

The function of a gene regulatory network depends on the architecture of the nodes , whose function is dependent on the multiple cis-regulatory modules. The layout of cis-regulatory modules can provide enough information to generate spatial and temporal patterns of gene expression. During development each domain, where each domain represents a different spatial regions of the embryo, of gene expression will be under the control of different cis-regulatory module(s). The design of regulatory modules help in producing feedback, feed forward, and cross-regulatory loops.

Genenetwork (database)

GeneNetwork is a database and open source bioinformatics software resource for systems genetics. This resource is used to study gene regulatory networks that link DNA sequence variants to corresponding differences in gene and protein expression and to differences in traits such as health and disease risk.

Data sets in GeneNetwork are typically made up of large collections of genotypes (e.g., SNPs) and phenotypes that are obtained from groups of related individuals, including human families, experimental crosses of strains of mice and rats, and organisms as diverse as Drosophila melanogaster, Arabidopsis thaliana, and barley.

The inclusion of genotypes for all individuals makes it practical to carry out web-based gene mapping to discover those regions of the genome that contribute to differences in gene expression, cell function, anatomy, physiology, and behavior among individuals.

History

GeneNetwork was created at the University of Tennessee Health Science Center, Memphis USA in 2000-2001. It was initially developed as a web-adapted version of Kenneth F.

Manly's Map Manager QT and QTX programs and was called WebQTL. Gene mapping data were incorporated for several mouse recombinant inbred strains. By early 2003, the first large Affymetrix gene expression data sets (whole mouse brain mRNA and hematopoietic stem cells) were incorporated and the system was renamed. GeneNetwork is now developed by an international group of developers and has mirror and development sites in Europe, Asia, and Australia. The production service is hosted on the Amazon Elastic Compute Cloud.

Organization and Use

GeneNetwork consists of two major components:

- Massive collections of genetic, genomic, and phenotype data for large families
- Sophisticated statistical analysis and gene mapping software that enable analysis of regulatory networks and genotype-to-phenotype relations

Four levels of data are usually obtained for each family or population:

1. DNA sequences and genotypes
2. Gene expression values using microarray, RNA-seq, or proteomic methods (molecular phenotypes)
3. Standard phenotypes of the type that are part of a typical medical record (e.g., blood chemistry, body weight)
4. Annotation files and metadata

The combined data types are housed together in a single relational database, but are conceptually organized and divided by species and family. The system is implemented as a LAMP (software bundle) stack. Code and a simplified version of the MySQL database are available at Sourceforge.net/projects/genenetwork/.

GeneNetwork is primarily used by researchers but has also been adopted successfully for undergraduate courses in genetics (see YouTube example), bioinformatics, physiology, and psychology. Researchers and students typically retrieve sets of genotypes and phenotypes from one or more families and use built-in statistical and mapping functions to explore relations among variables and to assemble networks of associations. Key steps include the analysis of these factors:

1. The range of variation of traits
2. Covariation among traits (scatterplots and correlations)
3. Architecture of larger networks of traits
4. Quantitative trait locus mapping and causal models of the linkage between sequence differences and phenotype differences

Data Sources

Massive expression data sets are submitted by researchers directly or are extracted from repositories such as National Center for Biotechnology Information Gene Expression Omnibus. A wide variety of cells and tissues are included—from single cell populations of the immune system, specific tissues (retina, prefrontal cortex), to entire systems (whole brain, lung, muscle, heart, fat, kidney, flower, even whole plant embryos). A typical data set is often based on hundreds of fully genotyped individuals and may also include biological replicates. Genotypes and phenotypes are taken from peer-reviewed papers. GeneNetwork includes annotation files for several RNA profiling platforms (Affymetrix, Illumina, and Agilent). RNA-seq data are also available for BXD recombinant inbred mice. Content and nomenclature are reviewed and edited by curators. Updates on coverage of species, families, tissues and measurement types are available at this site: .

Topics of annotation include the following:

- DNA sequence (SNPs, CNVs, indels)
- Transcriptomes (arrays, RNA-seq)
- Gene regulatory networks
- Phenome

Tools and Features

There are tools on the site for a wide range of functions that range from simple graphical displays of variation in gene expression or other phenotypes, scatter plots of pairs of traits (Pearson or rank order), construction of both simple and complex network graphs, analysis of principal components and synthetic traits, QTL mapping using marker regression, interval mapping, and pair scans for epistatic interactions. Most functions work with up to 100 traits and several functions work with an entire transcriptome.

The database can be browsed and searched at the main search page. An on-line tutorial is available. Users can also download the primary data sets as text files, Excel, or in the case of network graphs, as SBML.

Code

GeneNetwork is an open source project released under the Affero General Public License (AGPLv3). The majority of code is written in Python, but includes modules and other code written in C and JavaScript. GeneNetwork also calls statistical procedures written in the R programming language. The source code and a compact database are available on GeneNetwork sites and at SourceForge.

Morphogen

A morphogen is a substance governing the pattern of tissue development, and the positions of the various specialized cell types within a tissue. It spreads from a localized source and forms a concentration gradient across a developing tissue .

In developmental biology a morphogen is rigorously used to mean a signaling molecule that acts directly on cells (not through serial induction) to produce specific cellular responses dependent on morphogen concentration .

Well-known morphogens include: Decapentaplegic / Transforming growth factor beta, Hedgehog / Sonic Hedgehog, Wingless / Wnt, Epidermal growth factor, Fibroblast growth factor, and Retinoic acid.

Morphogens are defined conceptually, not chemically, so simple compounds such as retinoic acid (the active metabolite of retinol or vitamin A) may also act as morphogens

Introduction

During early development, morphogen gradients generate different cell types in distinct spatial order. The morphogen provides spatial information by forming a concentration gradient that subdivides a field of cells by inducing or maintaining the expression of different target genes at distinct concentration thresholds. Thus, cells far from the source of the morphogen will receive low levels of morphogen and express only low-threshold target genes.

In contrast, cells close to the source of morphogen will receive high levels of morphogen and will express both low- and high-threshold target genes. Distinct cell types emerge as a consequence of the different combination of target gene expression. In this way, the field of cells is subdivided into different types according to their position relative to the source of the morphogen.

This is a general mechanism by which cell type diversity can be generated in animal development.

Some of the earliest and best-studied morphogens are transcription factors that diffuse within early Drosophila melanogaster (fruit fly) embryos. However, most morphogens are secreted proteins that signal between cells.

Fruit Fly

Drosophila melanogaster has an unusual developmental system, in which the first thirteen cell divisions of the embryo occur within a syncytium prior to cellularization. Essentially the embryo remains a single cell with over 8000 nuclei evenly spaced near the membrane until the fourteenth cell division, when independent membranes furrow between the nuclei, separating them into independent cells. As a result, in fly embryos transcription factors such as Bicoid or Hunchback can act as morphogens because they can freely diffuse between nuclei to produce smooth gradients of concentration without relying on specialized intercellular signalling mechanisms. Although there is some evidence that homeobox transcription factors similar to these can pass directly through cell membranes, this mechanism is not believed to contribute greatly to morphogenesis in cellularized systems.

In most developmental systems, such as human embryos or later Drosophila development, syncytia occur only rarely (such as in skeletal muscle), and morphogens are generally secreted signalling proteins. These proteins bind to the extracellular domains of transmembrane receptor proteins, which use an elaborate process of signal transduction to communicate the level of morphogen to the nucleus. The nuclear targets of signal transduction pathways are usually transcription factors, whose activity is regulated in a manner that reflects the level of morphogen received at the cell surface. Thus, secreted morphogens act to generate gradients of transcription factor activity just like those that are generated in the syncitial Drosophila embryo.

Discrete target genes respond to different thresholds of morphogen activity. The expression of target genes is controlled by segments of DNA called 'enhancers' to which transcription factors bind directly. Once bound, the transcription factor then stimulates or inhibits the transcription of the gene and thus controls the level of expression of the gene product (usually a protein). 'Low-threshold' target genes require only low levels of morphogen activity to be regulated and feature enhancers that contain many high-affinity binding sites for the transcription factor. 'High-threshold' target genes have relatively fewer binding sites or low-affinity binding sites that require much greater levels of transcription factor activity to be regulated.

Thus, the general mechanism by which morphogens subdivide tissues into patterns of distinct cell types is well understood. However, morphogens often have additional activities such as controlling the growth of the tissue or orienting the polarity of cells within it (for example, the hairs on your forearm point in one direction). These activities of morphogens are much less understood and are the subject of current research efforts in the field of developmental biology.

Discovery

The morphogen idea has a long history in developmental biology, dating back to the work of the pioneering Drosophila geneticist, Thomas Hunt Morgan, in the early 20th century. However, it was Lewis Wolpert who refined the morphogen concept in the 1960s with his famous French flag model which described how morphogen could subdivide a tissue into domains of different target gene expression (corresponding to the colours of the French flag). This model was championed by the leading Drosophila biologist, Peter Lawrence. Christiane Nusslein-Volhard identified the first morphogen, Bicoid, one of the transcription factors present in a gradient in the Drosophila syncitial embryo. Two labs, that of Gary Struhl and that of Stephen Cohen, then demonstrated that a secreted signalling protein, Decapentaplegic (the Drosophila homologue of Transforming Growth Factor Beta), acted as a morphogen during later stages of Drosophila development.

Subsequent studies of the development of many different animals has confirmed the widespread importance of morphogens in governing animal development.

Operon

In genetics, an operon is a functioning unit of genomic DNA containing a cluster of genes under the control of a single regulatory signal or promoter. The genes are transcribed together into an mRNA strand and either translated together in the cytoplasm, or undergo trans-splicing to create monocistronic mRNAs that are translated separately, i.e. several strands of mRNA that each encode a single gene product. The result of this is that the genes contained in the operon are either expressed together or not at all. Several genes must be both co-transcribed and co-regulated to define an operon.

Originally, operons were thought to exist solely in prokaryotes, but since the discovery of the first operons in eukaryotes in the early 1990s, more evidence has arisen to suggest they are more common than previously assumed.

History

The term "operon" was first proposed in a short paper in the Proceedings of the French Academy of Science in 1960 . From this paper, the so-called general theory of the operon was developed. This theory suggested that all genes are controlled by means of operons through a single feedback regulatory mechanism– repression. Later, it was discovered that the regulation of genes is a much more complicated process. Indeed, it is not possible to talk of a general regulatory mechanism, because different operons have different mechanisms. Despite modifications, the development of the concept is considered a landmark event in the history of molecular biology. The first operon to be described was the lac operon in E. coli. The 1965 Nobel Prize in Physiology and Medicine was awarded to François Jacob, André Michel Lwoff and Jacques Monod for their discoveries concerning the operon and virus synthesis.

Overview

Operons occur primarily in prokaryotes but also in some eukaryotes, including nematodes such as C. elegans. , and Drosophila melanogaster flies. rRNA genes often exist in operons that have been found in a range of eukaryotes including chordates. An operon is made up of several structural genes arranged under a common promoter and regulated by a common operator. It is defined as a set of adjacent structural genes, plus the adjacent regulatory signals that affect transcription of the structural genes. The regulators of a given operon, including repressors, corepressors, and activators, are not necessarily coded for by that operon. The location and condition of the regulators, promoter, operator and structural DNA sequences can determine the effects of common mutations.

Operons are related to regulons, stimulons and modulons; whereas operons contain a set of genes regulated by the same operator, regulons contain a set of genes under regulation by a single regulatory protein, and stimulons contain a set of genes under regulation by a single cell stimulus. According to its authors, the term “operon” means “to operate”.

As a Unit of Transcription

An operon contains one or more structural genes which are generally transcribed into one polycistronic mRNA (a single mRNA molecule that codes for more than one protein). However, the definition of an operon does not require the mRNA to be polycistronic, though in practice, it usually is . Upstream of the structural genes lies a promoter sequence which provides a site for RNA polymerase to bind and initiate transcription. Close to the promoter lies a section of DNA called an operator. The operon may also contain regulatory genes such as a repressor gene, which codes for a regulatory protein that binds to the operator and inhibits transcription. Regulatory genes need not be part of the operon itself, but may be located elsewhere in the genome. The repressor molecule will reach the operator to block the transcription of the structural genes.

Structure

This is the general structure of an operon:

- Promoter – a nucleotide sequence that enables a gene to be transcribed. The promoter is recognized by RNA polymerase, which then initiates transcription. In RNA synthesis, promoters indicate which genes should be used for messenger RNA creation – and, by extension, control which proteins the cell manufactures.

- Operator – a segment of DNA that a regulator binds to. It is classically defined in the lac operon as a segment between the promoter and the genes of the operon. In the case of a repressor, the repressor protein physically obstructs the RNA polymerase from transcribing the genes.
- Structural genes – the genes that are co-regulated by the operon.

Regulation

Control of an operon is a type of gene regulation that enables organisms to regulate the expression of various genes depending on environmental conditions.

Operon regulation can be either negative or positive by induction or repression.

Negative control involves the binding of a repressor to the operator to prevent transcription.

- In negative inducible operons, a regulatory repressor protein is normally bound to the operator, which prevents the transcription of the genes on the operon. If an inducer molecule is present, it binds to the repressor and changes its conformation so that it is unable to bind to the operator. This allows for expression of the operon.
- In negative repressible operons, transcription of the operon normally takes place. Repressor proteins are produced by a regulator gene, but they are unable to bind to the operator in their normal conformation. However, certain molecules called corepressors are bound by the repressor protein, causing a conformational change to the active state. The activated repressor protein binds to the operator and prevents transcription.

Operons can also be positively controlled. With positive control, an activator protein stimulates transcription by binding to DNA (usually at a site other than the operator).

- In positive inducible operons, activator proteins are normally unable to bind to the pertinent DNA. When an inducer is bound by the activator protein, it undergoes a change in conformation so that it can bind to the DNA and activate transcription.
- In positive repressible operons, the activator proteins are normally bound to the pertinent DNA segment. However, when a corepressor is bound by the activator, it is prevented from binding the DNA. This stops activation and transcription of the system.

The Lac Operon

The lac operon of the model bacterium Escherichia coli was the first operon to be discovered and provides a typical example of operon function. It consists of three adjacent structural genes, a promoter, a terminator, and an operator. The lac operon is regulated by several factors including the availability of glucose and lactose. This is an example of the derepressible (from above: negative inducible) model.

The Trp Operon

Discovered in 1953 by Jacques Monod and colleagues, the trp operon in E. coli was the first repressible operon to be discovered. While the lac operon can be activated by a chemical (allolactose), the tryptophan (Trp) operon is inhibited by a chemical (tryptophan). This operon contains five structural genes: trp E, trp D, trp C, trp B, and trp A, which encodes tryptophan synthetase. It also contains a promoter which binds to RNA polymerase and an operator which blocks transcription when bound to the protein synthesized by the repressor gene (trp R) that binds to the operator. In the lac operon, lactose binds to the repressor protein and prevents it from repressing gene transcription, while in the trp operon, tryptophan binds to the repressor protein and enables it to repress gene transcription. Also unlike the lac operon, the trp operon contains a leader peptide and an attenuator sequence which allows for graded regulation. This is an example of the corepressible model.

Predicting the Number and Organization of Operons

The number and organization of operons has been studied most critically in E. coli. As a result, predictions can be made based on an organism's genomic sequence.

One prediction method uses the intergenic distance between reading frames as a primary predictor of the number of operons in the genome. The separation merely changes the frame and guarantees that the read through is efficient. Longer stretches exist where operons start and stop, often up to 40–50 bases.

An alternative method to predict operons is based on finding gene clusters where gene order and orientation is conserved in two or more genomes.

Operon prediction is even more accurate if the functional class of the molecules is considered. Bacteria have clustered their reading frames into units, sequestered by co-involvement in protein complexes, common pathways, or shared substrates and transporters. Thus, accurate prediction would involve all of these data, a difficult task indeed. Pascale Cossart published in 2009 the first full map of an operon, identifying the genetic switches that operate in Listeria under different conditions.

Synexpression

Synexpression is a type of non-random eukaryotic gene organization. Genes in a synexpression group may not be physically linked, but they are involved in the same process and they are coordinately expressed. It is expected that genes that function in the same process be regulated coordinately. Synexpression groups in particular represent genes that are simultaneously up- or down-regulated, often because their gene products are required in stoichiometric amounts or are protein-complex subunits. It is likely that these gene groups share common cis- and trans-acting control elements to achieve coordinate expression.

Synexpression groups are determined mainly by analysis of expression profiles compiled by the use of DNA microarrays. The use of this technology helps researchers monitor changes in expression patterns for large numbers of genes in a given experiment. Analysis of DNA microarray expression profiles has led to the discovery of a number of genes that are tightly co-regulated. One simplified example of a synexpression group is the genes cdc6, cdc3, cdc46, and swi4 in yeast, which are all co-expressed early in the G-1 stage of the cell cycle., These genes share one common cis-regulatory element, called ECB, which serves as a binding site for the MCM1 trans-acting protein. Although these genes are not spatially clustered, co-regulation seems to be achieved via this common cis and trans control mechanism. Most synexpression groups are more complicated than the ECB group in yeast, involving a myriad of cis and trans control elements.,

The identification of synexpression groups has had an impact on the way some scientists view evolutionary change in higher eukaryotes. Since groups of genes involved in the same biological process often share one or more common control elements, it has been suggested that the differential expression of these synexpression groups in different tissues of organisms can contribute to co-evolution tissues, organs, and appendages. Today it is commonly believed that it is not primarily the gene products themselves that evolve, but that it is the control networks for groups of genes that contribute most to the evolution of higher eukaryotes. Developmental processes provide an example of how changes in synexpression control networks could have a significant impact on an organism's capacity to evolve and adapt effectively. In animals, it is often beneficial for appendages to co-evolve, and it has been observed that fore-and hind-limbs share expression of Hox genes early in metazoan development. Thus, changes in the regulatory patterns of these genes would effect the development of both the fore- and hind-limbs, facilitating co-evolution.

Systems biology

Systems biology is a term used to describe a number of trends in bioscience research, and a movement which draws on those trends. Proponents describe systems biology as a biology-based inter-disciplinary study field that focuses on complex interactions in biological systems, claiming that it uses a new perspective (holism instead of reduction). Particularly from year 2000 onwards, the term is used widely in the biosciences, and in a variety of contexts. An often stated ambition of systems biology is the modeling and discovery of emergent properties, properties of a system whose theoretical description is only possible using techniques which fall under the remit of systems biology. These typically involve cell signaling networks, via long-range allostery.

Overview

Systems biology can be considered from a number of different aspects:

- As a field of study, particularly, the study of the interactions between the components of biological systems, and how these interactions give rise to the function and behavior of that system (for example, the enzymes and metabolites in a metabolic pathway).
- As a paradigm, usually defined in antithesis to the so-called reductionist paradigm (biological organisation), although fully consistent with the scientific method. The distinction between the two paradigms is referred to in these quotations:

"The reductionist approach has successfully identified most of the components and many of the interactions but, unfortunately, offers no convincing concepts or methods to understand how system properties emerge...the pluralism of causes and effects in biological networks is better addressed by observing, through quantitative measures, multiple components simultaneously and by rigorous data integration with mathematical models" Sauer et al "Systems biology...is about putting together rather than taking apart, integration rather than reduction. It requires that we develop ways of thinking about integration that are as rigorous as our reductionist programmes, but different....It means changing our philosophy, in the full sense of the term" Denis Noble

- As a series of operational protocols used for performing research, namely a cycle composed of theory, analytic or computational modelling to propose specific testable hypotheses about a biological system, experimental validation, and then using the newly acquired quantitative description of cells or cell processes to refine the computational model or theory. Since the objective is a model of the interactions in a system, the experimental techniques that most suit systems biology are those that are system-wide and attempt to be as complete as possible. Therefore, transcriptomics, metabolomics, proteomics and high-throughput techniques are used to collect quantitative data for the construction and validation of models.
- As the application of dynamical systems theory to molecular biology.
- As a socioscientific phenomenon defined by the strategy of pursuing integration of complex data about the interactions in biological systems from diverse experimental sources using interdisciplinary tools and personnel.

This variety of viewpoints is illustrative of the fact that systems biology refers to a cluster of peripherally overlapping concepts rather than a single well-delineated field. However the term has widespread currency and popularity as of 2007, with chairs and institutes of systems biology proliferating worldwide.

History

Systems biology finds its roots in:

- The quantitative modeling of enzyme kinetics, a discipline that flourished between 1900 and 1970,
- The mathematical modeling of population growth,
- The simulations developed to study neurophysiology, and
- Control theory and cybernetics.

One of the theorists who can be seen as one of the precursors of systems biology is Ludwig von Bertalanffy with his general systems theory. One of the first numerical simulations in biology was published in 1952 by the British neurophysiologists and Nobel prize winners Alan Lloyd Hodgkin and Andrew Fielding Huxley, who constructed a mathematical model that explained the action potential propagating along the axon of a neuronal cell. Their model described a cellular function emerging from the interaction between two different molecular components, a potassium and a sodium channel, and can therefore be seen as the beginning of computational systems biology. In 1960, Denis Noble developed the first computer model of the heart pacemaker. The formal study of systems biology, as a distinct discipline, was launched by systems theorist Mihajlo Mesarovic in 1966 with an international symposium at the Case Institute of Technology in Cleveland, Ohio entitled "Systems Theory and Biology". The 1960s and 1970s saw the development of several approaches to study complex molecular systems, such as the Metabolic Control Analysis and the biochemical systems theory. The successes of molecular biology throughout the 1980s, coupled with a skepticism toward theoretical biology, that then promised more than it achieved, caused the quantitative modelling of biological processes to become a somewhat minor field. However the birth of functional genomics in the 1990s meant that large quantities of high quality data became available, while the computing power exploded, making more realistic models possible. In 1997, the group of Masaru Tomita published the first quantitative model of the metabolism of a whole (hypothetical) cell. Around the year 2000, after Institutes of Systems Biology were established in Seattle and Tokyo, systems biology emerged as a movement in its own right, spurred on by the completion of various genome projects, the large increase in data from the omics (e.g. genomics and proteomics) and the accompanying advances in high-throughput experiments and bioinformatics.

Associated disciplines

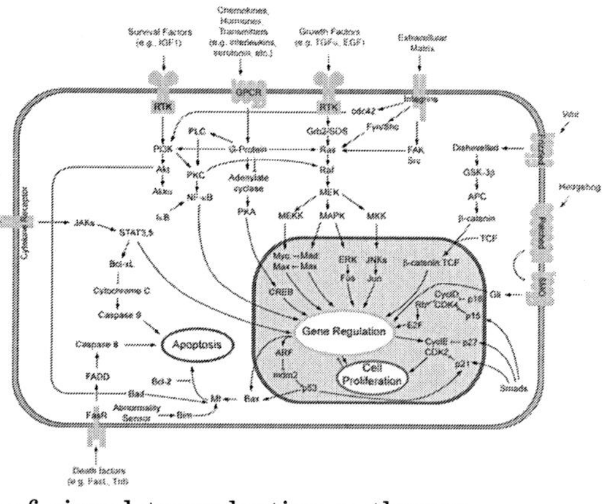

Figure:3 Overview of signal transduction pathways

According to the interpretation of Systems Biology as the ability to obtain, integrate and analyze complex data sets from multiple experimental sources using interdisciplinary tools, some typical technology platforms are:

- Phenomics: Organismal variation in phenotype as it changes during its life span.
- Genomics: Organismal deoxyribonucleic acid (DNA) sequence, including intra-organisamal cell specific variation. (i.e. Telomere length variation etc.).
- Epigenomics / Epigenetics: Organismal and corresponding cell specific transcriptomic regulating factors not empirically coded in the genomic sequence. (i.e. DNA methylation, Histone Acetelation etc.).
- Transcriptomics: Organismal, tissue or whole cell gene expression measurements by DNA microarrays or serial analysis of gene expression
- Interferomics: Organismal, tissue, or cell level transcript correcting factors (i.e. RNA interference)
- Translatomics / Proteomics: Organismal, tissue, or cell level measurements of proteins and peptides via two-dimensional gel electrophoresis, mass spectrometry or multi-dimensional protein identification techniques (advanced HPLC systems coupled with mass spectrometry). Sub disciplines include phosphoproteomics, glycoproteomics and other methods to detect chemically modified proteins.
- Metabolomics: Organismal, tissue, or cell level measurements of all small-molecules known as metabolites.
- Glycomics: Organismal, tissue, or cell level measurements of carbohydrates.
- Lipidomics: Organismal, tissue, or cell level measurements of lipids.

In addition to the identification and quantification of the above given molecules further techniques analyze the dynamics and interactions within a cell. This includes:

- Interactomics: Organismal, tissue, or cell level study of interactions between molecules. Currently the authoritative molecular discipline in this field of study is protein-protein interactions (PPI), although the working definition does not pre-clude inclusion of other molecular disciplines such as those defined here.
- NeuroElectroDynamics: Organismal, brain computing function as a dynamic system, underlying biophysical mechanisms and emerging computation by electrical interactions.

- Fluxomics: Organismal, tissue, or cell level measurements of molecular dynamic changes over time.
- Biomics: systems analysis of the biome.

The investigations are frequently combined with large-scale perturbation methods, including gene-based (RNAi, mis-expression of wild type and mutant genes) and chemical approaches using small molecule libraries. Robots and automated sensors enable such large-scale experimentation and data acquisition. These technologies are still emerging and many face problems that the larger the quantity of data produced, the lower the quality. A wide variety of quantitative scientists (computational biologists, statisticians, mathematicians, computer scientists, engineers, and physicists) are working to improve the quality of these approaches and to create, refine, and retest the models to accurately reflect observations.

The systems biology approach often involves the development of mechanistic models, such as the reconstruction of dynamic systems from the quantitative properties of their elementary building blocks. For instance, a cellular network can be modelled mathematically using methods coming from chemical kinetics and control theory. Due to the large number of parameters, variables and constraints in cellular networks, numerical and computational techniques are often used.

Other aspects of computer science and informatics are also used in systems biology. These include:

- New forms of computational model, such as the use of process calculi to model biological processes (notable approaches include stochastic ð-calculus, BioAmbients, Beta Binders, BioPEPA and Brane calculus) and constraint-based modeling.
- Integration of information from the literature, using techniques of information extraction and text mining.
- Development of online databases and repositories for sharing data and models, approaches to database integration and software interoperability via loose coupling of software, websites and databases, or commercial suits.
- Development of syntactically and semantically sound ways of representing biological models.

Up-regulation and Down-regulation

Up-regulation is a process that occurs within a cell triggered by a signal (originating internal or external to the cell), which results in increased expression of one or more genes and as a result the protein(s) encoded by those genes. On the converse, down-regulation is a process resulting in decreased gene and corresponding protein expression.

- Up-regulation occurs, for example, when a cell is deficient in some kind of receptor. In this case, more receptor protein is synthesized and transported to the membrane of the cell and, thus, the sensitivity of the cell is brought back to normal, reestablishing homeostasis.
- Down-regulation occurs, for example, when a cell is overstimulated by a neurotransmitter, hormone, or drug for a prolonged period of time, and the expression of the receptor protein is decreased in order to protect the cell .

Inducible vs. Repressible Systems

Gene Regulation can be summarized by the response of the respective system:

- Inducible systems - An inducible system is off unless there is the presence of some molecule (called an inducer) that allows for gene expression. The molecule is said to "induce expression". The manner by which this happens is dependent on the control mechanisms as well as differences between prokaryotic and eukaryotic cells.
- Repressible systems - A repressible system is on except in the presence of some molecule (called a corepressor) that suppresses gene expression. The molecule is said to "repress expression". The manner by which this happens is dependent on the control mechanisms as well as differences between prokaryotic and eukaryotic cells.

Theoretical Circuits

- Repressor/Inducer: an activation of a sensor results in the change of expression of a gene
- negative feedback: the gene product downregulates its own production directly or indirectly, which can result in
  - — Keeping transcript levels constant/proportional to a factor
  - — Inhibition of run-away reactions when coupled with a positive feedback loop
  - — Creating an oscillator by taking advantage in the time delay of transcription and translation, given that the mRNA and protein half-life is shorter

- Positive feedback: the gene product upregulates its own production directly or indirectly, which can result in
  - Signal amplification
  - Bistable switches when two genes inhibit each other and both have positive feedback
  - Pattern generation

Methods

In general, most experiments investigating differential expression used whole cell extracts of RNA, called steady-state levels, to determine which genes changed and by how much they did. These are, however, not informative of where the regulation has occurred and may actually mask conflicting regulatory processess but it is still the most commonly analysed (QPCR and DNA microarray).

When studying gene expression, there are several methods to look at the various stages. In eukaryotes these include:

- The local chromatin environment of the region can be determined by ChIP-chip analysis by pulling down RNA Polymerase II, Histone 3 modifications, Trithorax-group protein, Polycomb-group protein, or any other DNA-binding element to which a good antibody is available.
- Epistatic interactions can be investigated by synthetic genetic array analysis
- Due to post-transcriptional regulation, transcription rates and total RNA levels differ significantly. To measure the transcription rates nuclear run-on assays can be done and newer high-throughput methods are being developed, using thiol labelling instead of radioactivity.
- Only 5% of the RNA polymerised in the nucleus actually exists, and not only introns, abortive products, and non-sense transcripts are degradated. Therefore, the differences in nuclear and cytoplasmic levels can be see by separating the two fractions by gentle lysis.
- Alternative splicing can be analysed with a splicing array or with a tiling array.
- All in vivo RNA is complexed as RNPs. The quantity of transcripts bound to specific protein can be also analysed by RIP-Chip. For example, DCP2 will give an indication of sequestered protein; ribosome-bound gives and indication of transcripts active in

transcription (although it should be noted that a more dated method, called polysome fractionation, is still popular in some labs)

- Protein levels can be analysed by Mass spectrometry, which can be compared only to QPCR data, as microarray data is relative and not absolute.
- RNA and protein degradation rates are measured by means of transcription inhibitors (actinomycin D or á-amanitin) or translation inhibitors (Cycloheximide), respectively.

Nucleic Acid Methods

Nucleic acid methods are the techniques used to study nucleic acids (DNA and RNA).

Purification

- Phenol-chloroform extraction
- minicolumn purification
- RNA extraction

Quantification

- Abundance in weight: spectroscopic quantification
- Absolute abundance in number: Q-PCR
- high-throughput relative abundance DNA microarray
- high-throughput absolute abundance SAGE
- Size: Gel electrophoresis

Synthesis

- De novo: Oligonucleotide synthesis
- Amplification: PCR

Kinetics

- Dual polarization interferometry
- QCM-D

Other

- Nucleic acid simulations
- DNA sequencing
- Bisulfite sequencing
- Expression cloning
- Southern blot
- Northern blot

- Sucrose gradient centrifugation
- Radioactivity in biological research
- Lab-on-a-chip
- Nuclear run-on assay
- Fluorescent in situ hybridization
- several Bioinformatics methods, such as RNA structure prediction

Protein Methods

Protein methods are the techniques used to study proteins.

There are genetic methods for studying proteins, methods for detecting proteins, methods for isolating and purifying proteins and other methods for characterizing the structure and function of proteins, often requiring that the protein first be purified.

Genetic Methods

- Conceptual translation- many proteins are never directly sequenced, but their sequence of amino acids is known by "conceptual translation" of a known mRNA sequence. See genetic code.
- Site-directed mutagenesis allows new variants of proteins to be produced and tested for how structural changes alter protein function.
  — Insertion of protein tags such as the His-tag. See also green fluorescent protein.
- Evolutionary; analysis of sequence changes in different species using software such as BLAST.
- Proteins that are involved in human diseases can be identified by matching alleles to disease and other phenotypes using methods such as calculation of LOD scores.

Detecting Proteins

- microscopy, protein immunostaining
- Protein immunoprecipitation
- Immunoelectrophoresis
- Immunoblotting
- BCA Protein Assay
- Western blot
- Spectrophotometry
- Enzyme assay

Protein Purification

- Protein Isolation
  - Chromatography methods: Ion exchange, Size-exclusion chromatography or Gelfiltration, Affinity chromatography
- Protein Extraction and Solubilization
- Protein Concentration Determination Methods, Bradford protein assay
- Concentrating Protein Solutions
- Gel electrophoresis
  - Gel Electrophoresis Under denaturing conditions
  - Gel Electrophoresis Under non-denaturing conditions
  - 2D Gel Electrophoresis
- Electrofocusing

Protein Structures

- X-ray crystallography
- Protein NMR

Protein-DNA Interactions

- ChIP-on-chip
- Chip-Sequencing
- DamID
- Microscale Thermophoresis

Other Methods

- Hydrogen-deuterium exchange
- Mass spectrometry
- Molecular dynamics
- Protein structure prediction
- Protein sequencing
- Protein structural alignment
- Protein ontology
- Protein synthesis
- Proteomics
- Peptide mass fingerprinting
- Ligand binding assay
- Eastern blotting

- metabolic labeling
  - heavy isotope labeling
  - radioactive isotope labeling
- Enhancer (genetics)

In genetics, an enhancer is a short region of DNA that can be bound with proteins (namely, the trans-acting factors, much like a set of transcription factors) to enhance transcription levels of genes (hence the name) in a gene cluster. While enhancers are usually cis-acting, an enhancer does not need to be particularly close to the genes it acts on, and sometimes need not be located on the same chromosome. Enhancer locations In eukaryotic cells the structure of the chromatin complex of DNA is folded in a way that functionally mimics the supercoiled state characteristic of prokaryotic DNA, so that although the enhancer DNA is far from the gene in regard to the number of nucleotides, it is geometrically close to the promoter and gene. This allows it to interact with the general transcription factors and RNA polymerase II.

An enhancer may be located upstream or downstream of the gene it regulates. Furthermore, an enhancer need not be located near to the transcription initiation site to affect transcription, as some have been found located in several hundred thousand base pairs upstream or downstream of the start site.

Enhancers do not act on the promoter region itself, but are bound by activator proteins. These activator proteins interact with the mediator complex, which recruits polymerase II and the general transcription factors which then begin transcribing the genes. Enhancers can also be found within introns. An enhancer's orientation may even be reversed without affecting its function. Additionally, an enhancer may be excised and inserted elsewhere in the chromosome, and still affect gene transcription. That is one reason that intron polymorphisms may have effects although they are not translated.

Theories

Currently, there are two different theories on the information processing that occurs on enhancers:

- Enhanceosomes - rely on highly cooperative, coordinated action and can be disabled by single point mutations that move or remove the binding sites of individual proteins.
- Flexible billboards - less integrative, multiple proteins independently regulate gene expression and their sum is read in by the basal transcriptional machinery

HACNS1

HACNS1 (also known as CENTG2 and located in the Human Accelerated Region 2) is a gene enhancer "that may have contributed to the evolution of the uniquely opposable human thumb, and possibly also modifications in the ankle or foot that allow humans to walk on two legs". Evidence to date shows that of the 110,000 gene enhancer sequences identified in the human genome, HACNS1 has undergone the most change during the evolution of humans following the split with the ancestors of chimpanzees.

GADD45G

An enhancer near the gene GADD45g has been described that may regulate brain growth in chimpanzees and other mammals, but not in humans. The GADD45G regulator in mice and chimps is active in regions of the brain where cells that form the cortex, ventral forebrain, and thalamus are located and may suppress further neurogenesis. Loss of the GADD45G enhancer in humans may contribute to an increase of certain neuronal populations and to forebrain expansion in humans.

Enhancers in Developmental Biology

The development, differentiation and growth of cells and tissues require precisely regulated patterns of gene expression. Enhancers work as cis-regulatory elements to mediate both spatial and temporal control of development by turning on transcription in specific cells and/or repressing it in other cells. Thus, the particular combination of transcription factors and other DNA-binding proteins in a developing tissue controls which genes will be expressed in that tissue. Enhancers allow the same gene to be used in diverse processes in space and time.

Identification and Characterization

Enhancers can be identified by enhancer trap techniques using a reporter gene or by comparative sequence analysis and computational genomics. In genetically tractable models such as the fruit fly Drosophila melanogaster, for example, a reporter construct such as the lacZ gene can be randomly integrated into the genome using a P element transposon. If the reporter gene integrates near an enhancer, its expression will reflect the expression pattern driven by that enhancer. Thus, staining the flies for LacZ expression or activity and cloning the sequence surrounding the integration site can identify the enhancer sequence. In the comparative genomics approach, sequence conservation of non-coding regions can be indicative of enhancers. Sequences from multiple species are aligned, and conserved regions are identified computationally. Identified sequences can then be attached to a reporter gene such green fluorescent protein or lacZ to determine the in vivo pattern of gene expression produced by the enhancer when injected into an embryo. mRNA expression of the reporter can be visualized by in situ hybridization, which provides a more direct measure of enhancer activity, since it is not subjected to the complexities of translation and protein folding.

Although much evidence has pointed to sequence conservation for critical developmental enhancers, other work has shown that the function of enhancers can be conserved with little or no primary sequence conservation. For example, the RET enhancers in humans have very little sequence conservation to those in zebrafish, yet both species' sequences produce nearly identical patterns of reporter gene expression in zebrafish. Additionally, algorithms such as TFSEARCH or JASPAR can be used to identify putative transcription factor binding sites to further characterize an enhancer sequence.

Enhancers in Invertebrate Segmentation

The enhancers determining early segmentation in Drosophila melanogaster embryos are among the best characterized developmental enhancers. In the early fly embryo, the gap gene transcription factors are responsible for activating and repressing a number of segmentation genes, such as the pair rule genes.

The gap genes are expressed in blocks along the anterior-posterior axis of the fly along with other maternal effect transcription factors, thus creating zones within which different combinations of transcription factors are expressed. The pair-rule genes are separated from one another by non-expressing cells. Moreover, the stripes of expression for different pair-rule genes are offset by a few cell diameters from one another. Thus, unique combinations of pair-rule gene expression create spatial domains along the anterior-posterior axis to set up each of the 14 individual segments.

The 480 bp enhancer responsible for driving the sharp stripe two of the pair-rule gene even-skipped (eve) has been well-characterized. The enhancer contains 12 different binding sites for maternal and gap gene transcription factors. Activating and repressing sites overlap in sequence. Eve is only expressed in a narrow stripe of cells that contain high concentrations of the activators and low concentration of the repressors for this enhancer sequence. Other enhancer regions drive eve expression in 6 other stripes in the embryo.

Enhancers in Vertebrate Patterning

Establishing body axes is a critical step in animal development. During mouse embryonic development, Nodal, a transforming growth factor-beta superfamily ligand, is a key gene involved in patterning both the anterior-posterior axis and the left-right axis of the early embryo. The Nodal gene contains two enhancers: the Proximal Epiblast Enhancer (PEE) and the Asymmetric Enhancer (ASE). The PEE is upstream of the Nodal gene and drives Nodal expression in the portion of the primitive streak that will differentiate into the node (also referred to as the primitive node). The PEE turns on Nodal expression in response to a combination of Wnt signaling plus a second, unknown signal; thus, a member of the LEF/TCF transcription factor family likely binds to a TCF binding site in the cells in the node. Diffusion of Nodal away from the node forms a gradient which then patterns the extending anterior-posterior axis of the embryo. The ASE is an intronic enhancer bound by the fork head domain transcription factor Fox1. Early in development, Fox1-driven Nodal expression establishes the visceral endoderm. Later in development, Fox1 binding to the ASE drives Nodal expression on the left side of the lateral plate mesoderm, thus establishing left-right asymmetry necessary for asymmetric organ development in the mesoderm.

Establishing three germ layers during gastrulation is another critical step in animal development. Each of the three germ layers has unique patterns of gene expression that promote their differentiation and development. The endoderm is specified early in development by Gata4 expression, and Gata4 goes on to direct gut morphogenesis later. Gata4 expression is controlled in the early embryo by an intronic enhancer that binds another forkhead domain transcription factor, FoxA2. Initially the enhancer drives broad gene expression throughout the embryo, but the expression quickly becomes restricted to the endoderm, suggesting that other repressors may be involved in its restriction. Late in development, the same enhancer restricts expression to the tissues that will become the stomach and pancreas. An additional enhancer is responsible for maintaining Gata4 expression in the endoderm during the intermediate stages of gut development.

Multiple Enhancers Promote Developmental Robustness

Some genes involved in critical developmental processes contain multiple enhancers of overlapping function. Secondary enhancers, or "shadow enhancers," may be found many kilobases away from the primary enhancer ("primary" usually refers to the first enhancer discovered, which is often closer to the gene it regulates). On its own, each enhancer drives nearly identical patterns of gene expression. Are the two enhancers truly redundant? Recent work has shown that multiple enhancers allow fruit flies to survive environmental perturbations, such as an increase in temperature. When raised at an elevated temperature, a single enhancer sometimes fails to drive the complete pattern of expression, whereas the presence of both enhancers permits normal gene expression.

Enhancers and the Evolution of Developmental Mechanisms

One theme of research in evolutionary developmental biology ("evo-devo") is investigating the role of enhancers and other cis-regulatory elements in producing morphological changes via developmental differences between species.

Stickleback PitX1

Recent work has investigated the role of enhancers in morphological changes in threespine stickleback fish. Sticklebacks exist in both marine and freshwater environments, but sticklebacks in many freshwater populations have completely lost their pelvic fins (appendages homologous to the posterior limb of tetrapods). Pitx1 is a homeobox gene involved in posterior limb development in vertebrates. Preliminary genetic analyses indicated that changes in the expression of this gene were responsible for pelvic reduction in sticklebacks. Fish expressing only the freshwater allele of Pitx1 do not have pelvic spines, whereas fish expressing a marine allele retain pelvic spines. A more thorough characterization showed that a 500 base pair enhancer sequence is responsible for turning on Pitx1 expression in the posterior fin bud. This enhancer is located near a chromosomal fragile site—a sequence of DNA that is likely to be broken and thus more likely to be mutated as a result of imprecise DNA repair. This fragile site has caused repeated, independent losses of the enhancer responsible for driving PitX1 expression in the pelvic spines in isolated freshwater population, and without this enhancer, freshwater fish fail to develop pelvic spines.

Enhancers in Drosophila wing Pattern Evolution

Pigmentation patterns provide one of the most striking and easily scored differences between different species of animals. Pigmentation of the Drosophila wing has proven to be a particularly amenable system for studying the development of complex pigmentation phenotypes. The Drosophila guttifera wing has 12 dark pigmentation spots and 4 lighter gray intervein patches. Pigment spots arise from expression of the yellow gene, whose product produces black melanin. Recent work has shown that two enhancers in the yellow gene produce gene expression in precisely this pattern – the vein spot enhancer drives reporter gene expression in the 12 spots, and the intervein shade enhancer drives reporter expression in the 4 distinct patches. These two enhancers are responsive to the Wnt signaling pathway, which is activated by "wingless" expression at all of the pigmented locations. Thus, in the evolution of the complex pigmentation phenotype, the yellow pigment gene evolved enhancers responsive to the wingless signal and wingless expression evolved at new locations to produce novel wing patterns.

- Artificial transcription factors (small molecules that mimic transcription factor protein)

An artificial transcription factor is an example of a chimeral protein, designed to target and modulate gene transcription.

They are generally composed of a DNA-binding domain (specific to a certain sequence) coupled to a modulatory domain (which acts upon other transcription factors) in order to alter the expression of a particular gene. It is also possible to downregulate expression of a gene by targeting the 5' untranslated region with a DNA-binding domain that lacks a regulatory domain; this will reduce transcription simply by blocking RNA polymerase progression along the DNA template.

Cellular model

Creating a cellular model has been a particularly challenging task of systems biology and mathematical biology. It involves developing efficient algorithms, data structures, visualization and communication tools to orchestrate the integration of large quantities of biological data with the goal of computer modeling. It is also directly associated with bioinformatics, computational biology and Artificial life.

It involves the use of computer simulations of the many cellular subsystems such as the networks of metabolites and enzymes which comprise metabolism, signal transduction pathways and gene regulatory networks to both analyze and visualize the complex connections of these cellular processes. The complex network of biochemical reaction/transport processes and their spatial organization make the development of a predictive model of a living cell a grand challenge for the 21st century.

Overview

The eukaryotic cell cycle is very complex and is one of the most studied topics, since its misregulation leads to cancers. It is possibly a good example of a mathematical model as it deals with simple calculus but gives valid results. Two research groups have produced several models of the cell cycle simulating several organisms. They have recently produced a generic eukaryotic cell cycle model which can represent a particular eukaryote depending on the values of the parameters, demonstrating that the idiosyncrasies of the individual cell cycles are due to different protein concentrations and affinities, while the underlying mechanisms are conserved (Csikasz-Nagy et al., 2006).

By means of a system of ordinary differential equations these models show the change in time (dynamical system) of the protein inside a single typical cell; this type of model is called a deterministic process (whereas a model describing a statistical distribution of protein concentrations in a population of cells is called a stochastic process).

To obtain these equations an iterative series of steps must be done: first the several models and observations are combined to form a consensus diagram and the appropriate kinetic laws are chosen to write the differential equations, such as rate kinetics for stoichiometric reactions, Michaelis-Menten kinetics for enzyme substrate reactions and Goldbeter–Koshland kinetics for ultrasensitive transcription factors, afterwards the parameters of the equations (rate constants, enzyme efficiency coefficients and Michealis constants) must be fitted to match observations; when they cannot be fitted the kinetic equation is revised and when that is not possible the wiring diagram is modified. The parameters are fitted and validated using observations of both wild type and mutants, such as protein half-life and cell size.

In order to fit the parameters the differential equations need to be studied. This can be done either by simulation or by analysis. In a simulation, given a starting vector (list of the values of the variables), the progression of the system is calculated by solving the equations at each time-frame in small increments. In analysis, the proprieties of the equations are used to investigate the behavior of the system depending of the values of the parameters and variables. A system of differential equations can be represented as a vector field, where each vector described the change (in concentration of two or more protein) determining where and how fast the trajectory (simulation) is heading. Vector fields can have several special points: a stable point, called a sink, that attracts in all directions (forcing the concentrations to be at a certain value), an unstable point, either a source or a saddle point which repels (forcing the concentrations to change away from a certain value), and a limit cycle, a closed trajectory towards which several trajectories spiral towards (making the concentrations oscillate). A better representation which can handle the large number of variables and parameters is called a bifurcation diagram (bifurcation theory): the presence of these special steady-state points at certain values of a parameter (e.g. mass) is represented by a point and once the parameter passes a certain value, a qualitative change occurs, called a bifurcation, in which the nature of the space changes, with profound consequences for the protein concentrations:

The cell cycle has phases (partially corresponding to G1 and G2) in which mass, via a stable point, controls cyclin levels, and phases (S and M phases) in which the concentrations change independently, but once the phase has changed at a bifurcation event (cell cycle checkpoint), the system cannot go back to the previous levels since at the current mass the vector field is profoundly different and the mass cannot be reversed back through the bifurcation event, making a checkpoint irreversible.

In particular the S and M checkpoints are regulated by means of special bifurcations called a Hopf bifurcation and an infinite period bifurcation.

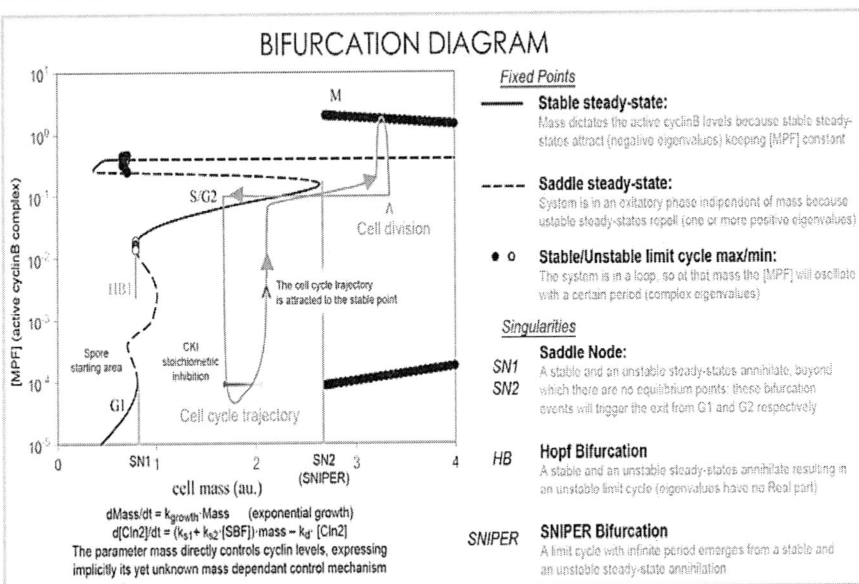

Molecular Level Simulations

E-Cell Project aims "to make precise whole cell simulation at the molecular level possible".

Projects

Multiple projects are in progress.

- Karyote - Indiana University
- E-Cell Project
- Virtual Cell - University of Connecticut Health Center
- Silicon Cell

Biological Data Visualization

Biology Data Visualization is a branch of bioinformatics concerned with the application of computer graphics, scientific visualization, and information visualization to different areas of the life sciences. This includes visualization of sequences, genomes, alignments, phylogenies, macromolecular structures, systems biology, microscopy, and magnetic resonance imaging data. Software tools used for visualizing biological data range from simple, stand-alone programs to complex, integrated systems.

State-of-the-art and Perspectives

Today we are experiencing a rapid growth in volume and diversity of biological data, presenting an increasing challenge for biologists. A key step in understanding and learning from these data is visualization. Thus, there has been a corresponding increase in the number and diversity of systems for visualizing biological data. An emerging trend is the blurring of boundaries between the visualization of 3D structures at atomic resolution, visualization of larger complexes by cryo-electron microscopy, and visualization of the location of proteins and complexes within whole cells and tissues. A second emerging trend is an increase in the availability and importance of time-resolved data from systems biology, electron microscopy and cell and tissue imaging. In contrast, visualization of trajectories has long been a prominent part of molecular dynamics. Finally, as datasets are increasing in size, complexity, and intercon-nectness, biological visualization systems are improving in usability, data integration and standardization.

List of Visualization Systems

A large number of software systems are available for visualization biological data. The links below link lists of such systems, grouped by application areas.

Conserved non-coding DNA sequence

A conserved non-coding sequence (CNS) is a DNA sequence of noncoding DNA that is evolutionarily conserved. These sequences are of interest for their potential to regulate gene production.

CNSs in plants and animals are highly associated with transcription factor binding sites and other cis-acting regulatory elements. Conserved non-coding sequences can be important sites of evolutionary divergence as mutations in these regions may alter the regulation of conserved genes, producing species-specific patterns of gene expression.

Spatiotemporal gene expression

Spatiotemporal gene expression is the activation of genes within specific tissues of an organism at specific times during development. Gene activation patterns vary widely in complexity. Some are straightforward and static, such as the pattern of tubulin, which is expressed in all cells at all times in life. Some, on the other hand, are extraordinarily intricate and difficult to predict and model, with expression fluctuating wildly from minute to minute or from cell to cell. Spatiotemporal variation plays a key role in generating the diversity of cell types found in developed organisms; since the identity of a cell is specified by the collection of genes actively expressed within that cell, if gene expression was uniform spatially and temporally, there could be at most one kind of cell.

Consider the gene wingless, a member of the wnt family of genes. In the early embryonic development of the model organism Drosophila melanogaster, or fruit fly, wingless is expressed across almost the entire embryo in alternating stripes three cells separated. This pattern is lost by the time the organism develops into a larva, but wingless is still expressed in a variety of tissues such as the wing imaginal discs, patches of tissue that will develop into the adult wings. The spatiotemporal pattern of wingless gene expression is determined by a network of regulatory interactions consisting of the effects of many different genes such as even-skipped and Krüppel.

What causes spatial and temporal differences in the expression of a single gene? Because current expression patterns depend strictly on previous expression patterns, there is a regressive problem of explaining what caused the first differences in gene expression. The process by which uniform gene expression becomes spatially and temporally differential is known as symmetry breaking. For example, in the case of embryonic Drosophila development, the genes nanos and bicoid are asymmetrically expressed in the oocyte because maternal cells deposit messenger RNA (mRNA) for these genes in the poles of the egg before it is laid.

Identifying Spatiotemporal Patterns

One way to identify the expression pattern of a particular gene is to place a reporter gene downstream of its promoter. In this configuration, the promoter gene will cause the reporter gene to be expressed only where and when the gene of interest is expressed. The expression distribution of the reporter gene can be determined by visualizing it. For example, the reporter gene green fluorescent protein can be visualized by stimulating it with blue light and then using a digital camera to record green fluorescent emission.

If the promoter of the gene of interest is unknown, there are several ways to identify its spatiotemporal distribution. Immunohistochemistry involves preparing an antibody with specific affinity for the protein associated with the gene of interest. This distribution of this antibody can then be visualized by a technique such as fluorescent labeling. Immunohistochemistry has the advantages of being methodologically feasible and relatively inexpensive. Its disadvantages include non-specificity of the antibody leading to false positive identification of expression. Poor penetrance of the antibody into the target tissue can lead to false negative results. Furthermore, since immunohistoc-hemistry visualizes the protein generated by the gene, if the protein product diffuses between cells, or has a particularly short or long half-life relative to the mRNA that is used to translate the protein, this can lead to distorted interpretation of which cells are expressing the mRNA.

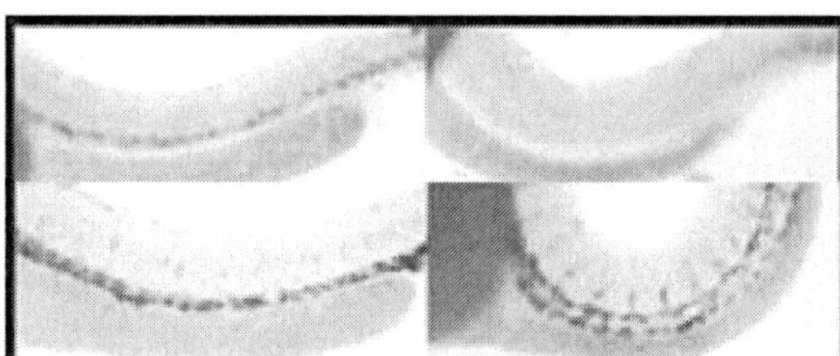

In situ hybridizations of genes expressed in arteries (top) and veins (bottom) in zebrafish. Blue staining indicates presence of the gene mRNAs. Panels on the left are normal animals, while animals on the right are mutated in the Notch gene. Fish lacking Notch have fewer arteries and more veins at this point in developmental time.

In situ hybridization is an alternate method in which a "probe," a synthetic nucleic acid with a sequence complementary to the mRNA of the gene, is added to the tissue. This probe is then chemically tagged so that it can be visualized later. This technique enables visualization specifically of mRNA-producing cells without any of the artifacts associated with immunohistochemistry. However, it is notoriously difficult, and requires knowledge of the sequence of DNA corresponding to the gene of interest. A method called enhancer-trap screening reveals the diversity of spatiotemporal gene expression patterns possible in an organism. In this technique, DNA that encodes a reporter gene is randomly inserted into the genome. Depending on the gene promoters proximal to the insertion point, the reporter gene will be expressed in particular tissues at particular points in development. While enhancer-trap derived expression patterns do not necessarily reflect the actual patterns of expression of specific genes, they reveal the variety of spatiotemporal patterns that are accessible to evolution.

Reporter genes can be visualized in living organisms, but both immunohistochemistry and in situ hybridization must be performed in fixed tissues. Techniques that require fixation of tissue can only generate a single temporal time point per individual organism. However, using live animals instead of fixed tissue can be crucial in dynamically understanding expression patterns over an individual's lifespan. Either way, variation between individuals can confound the interpretation of temporal expression patterns.

Post-transcriptional Regulation

In eukaryotes, where export of RNA is required before translation is possible, nuclear export is thought to provide additional control over gene expression. All transport in and out of the nucleus is via the nuclear pore and transport is controlled by a wide range of importin and exportin proteins.

Expression of a gene coding for a protein is only possible if the messenger RNA carrying the code survives long enough to be translated. In a typical cell an RNA molecule is only stable if specifically protected from degradation. RNA degradation has particular importance in regulation of expression in eukaryotic cells where mRNA has to travel significant distances before being translated. In eukaryotes RNA is stabilised by certain post-transcriptional modifications, particularly the 5' cap and poly-adenylated tail.

Intentional degradation of mRNA is used not just as a defence mechanism from foreign RNA (normally from viruses) but also as a route of mRNA destabilisation. If an mRNA molecule has a complementary sequence to a small interfering RNA then it is targeted for destruction via the RNA interference pathway.

Translational Regulation

Neomycin is an example of a small molecule which reduces expression of all protein genes inevitably leading to cell death, thus acts as an antibiotic. Direct regulation of translation is less prevalent than control of transcription or mRNA stability but is occasionally used. Inhibition of protein translation is a major target for toxins and antibiotics in order to kill a cell by overriding its normal gene expression control. Protein synthesis inhibitors include the antibiotic neomycin and the toxin ricin.

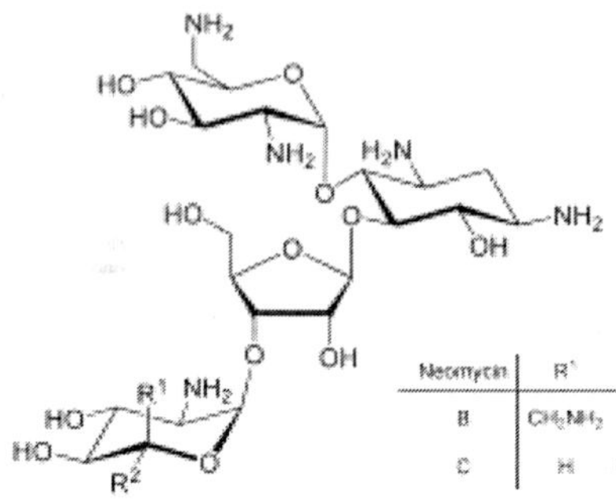

Neomycin	R1	R2
B	CH_2NH_2	H
C	H	CH_2NH_2

Protein Degradation

Once protein synthesis is complete the level of expression of that protein can be reduced by protein degradation. There are major protein degradation pathways in all prokaryotes and eukaryotes of which the proteasome is a common component. An unneeded or damaged protein is often labelled for degradation by addition of ubiquitin.

Measurement

Measuring gene expression is an important part of many life sciences - the ability to quantify the level at which a particular gene is expressed within a cell, tissue or organism can give a huge amount of information. For example measuring gene expression can:

- Identify viral infection of a cell (viral protein expression)
- Determine an individual's susceptibility to cancer (oncogene expression)
- Find if a bacterium is resistant to penicillin (beta-lactamase expression)

Similarly the analysis of the location of expression protein is a powerful tool and this can be done on an organism or cellular scale. Investigation of localisation is particularly important for study of development in multicellular organisms and as an indicator of protein function in single cells. Ideally measurement of expression is done by detecting the final gene product (for many genes this is the protein) however it is often easier to detect one of the precursors, typically mRNA, and infer gene expression level.

mRNA Quantification

Levels of mRNA can be quantitatively measured by northern blotting which gives size and sequence information about the mRNA molecules. A sample of RNA is separated on an agarose gel and hybridized to a radio-labeled RNA probe that is complementary to the target sequence. The radio-labeled RNA is then detected by an autoradiograph. The main problems with Northern blotting stem from the use of radioactive reagents (which make the procedure time consuming and potentially dangerous) and lower quality quantification than more modern methods (due to the fact that quantification is done by measuring band strength in an image of a gel). Northern blotting is, however, still widely used as the additional mRNA size information allows the discrimination of alternately spliced transcripts.

A more modern low-throughput approach for measuring mRNA abundance is reverse transcription quantitative polymerase chain reaction (RT-PCR followed with qPCR). RT-PCR first generates a DNA template from the mRNA by reverse transcription, which is called cDNA. This cDNA template is then used for qPCR where the change in fluorescence of a probe changes as the DNA amplification process progresses.

With a carefully constructed standard curve qPCR can produce an absolute measmsensitive (detection of a single mRNA molecule is possible), but can be expensive due to the fluorescent probes required.

An even more advanced approach is to individually tag single mRNA molecules with fluorescent barcodes (nanostrings), which can be detected one-by-one and counted for direct digital quantification. The advantage of this approach is that it does not rely on analog quantification of fluorescent intensity, which can be problematic due to noise, lack of linearity, and narrow dynamic range. Instead, the technique relies on fluorescence to detect simply the presence of a single mRNA molecule in a binary ("yes" or "no") mode. This method was invented by Dr. Krassen Dimitrov at the Institute for Systems Biology and commercialized through his start-up company, NanoString Technologies

Northern blots and RT-qPCR are good for detecting whether a single gene is being expressed, but it quickly becomes impractical if many genes within the sample are being studied. Using DNA microarrays, transcript levels for many genes at once (expression profiling) can be measured. Recent advances in microarray technology allow for the quantification, on a single array, of transcript levels for every known gene in several organism's genomes, including humans.

Alternatively "tag based" technologies like Serial analysis of gene expression (SAGE), which can provide a relative measure of the cellular concentration of different mRNAs, can be used. The great advantage of tag-based methods is the "open architecture", allowing for the exact measurement of any transcript, with a known or unknown sequence.

Protein Quantification

For genes encoding proteins the expression level can be directly assessed by a number of means with some clear analogies to the techniques for mRNA quantification.

The most commonly used method is to perform a Western blot against the protein of interest - this gives information on the size of the protein in addition to its identity. A sample (often cellular lysate) is separated on a polyacrylamide gel, transferred to a membrane and then probed with an antibody to the protein of interest. The antibody can either be conjugated to a fluorophore or to horseradish peroxidase for imaging and/or quantification. The gel-based nature of this assay makes quantification less accurate but it has the advantage of being able to identify later modifications to the protein, for example proteolysis or ubiquitination, from changes in size.

Localisation

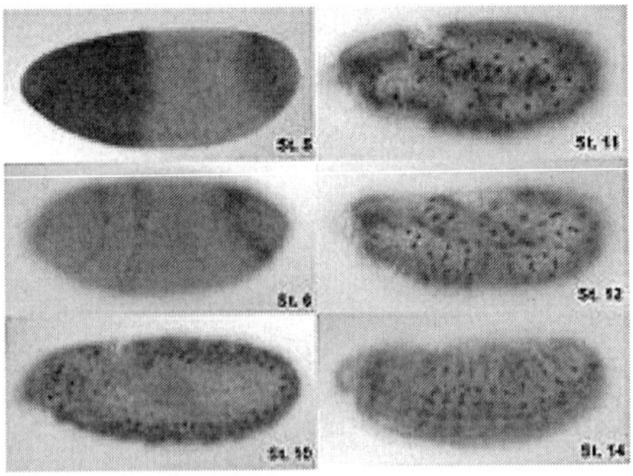

In situ-hybridization of Drosophila embryos at different developmental stages for the mRNA responsible for the expression of hunchback. High intensity of blue color marks places with high hunchback mRNA quantity. Analysis of expression is not limited to only quantification; localisation can also be determined. mRNA can be detected with a suitably labelled complementary mRNA strand and protein can be detected via labelled antibodies. The probed sample is then observed by microscopy to identify where the mRNA or protein is.

By replacing the gene with a new version fused a green fluorescent protein (or similar) marker expression may be directly quantified in live cells. This is done by imaging using a fluorescence microscope. It is very difficult to clone a GFP-fused protein into its native location in the genome without affecting expression levels so this method often cannot be used to measure endogenous gene expression. It is, however, widely used to measure the expression of a gene artificially introduced into the cell, for example via an expression vector. It is important to note that by fusing a target protein to a fluorescent reporter the protein's behavior, including its cellular localization and expression level, can be significantly changed.

The enzyme-linked immunosorbent assay works by using antibodies immobilised on a microtiter plate to capture proteins of interest from samples added to the well. Using a detection antibody conjugated to an enzyme or fluorophore the quantity of bound protein can be accurately measured by fluorometric or colourimetric detection. The detection process is very similar to that of a Western blot, but by avoiding the gel steps moreaccurate quantification can be achieved.

Expression System

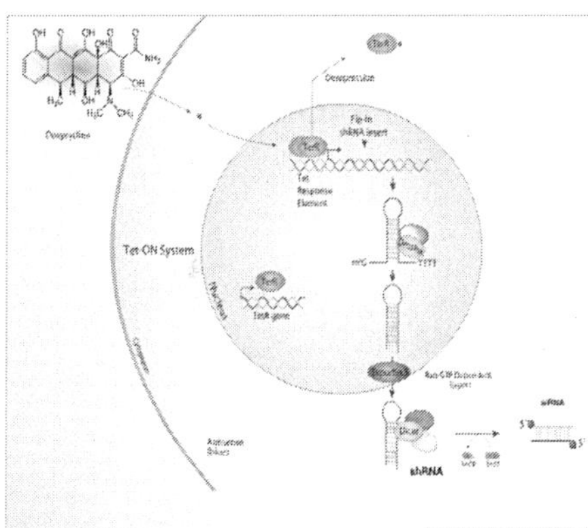

Figure: Tet-ON inducible shRNA system

An expression system is a system specifically designed for the production of a gene product of choice. This is normally a protein although may also be RNA, such as tRNA or a ribozyme. An expression system consists of a gene, normally encoded by DNA, and the molecular machinery required to transcribe the DNA into mRNA and translate the mRNA into protein using the reagents provided. In the broadest sense this includes every living cell but the term is more normally used to refer to expression as a laboratory tool. An expression system is therefore often artificial in some manner. Expression systems are, however, a fundamentally natural process. Viruses are an excellent example where they replicate by using the host cell as an expression system for the viral proteins and genome.

Inducible Expression

Doxycycline is also used in "Tet-on" and "Tet-off" tetracycline controlled transcriptional activation to regulate transgene expression in organisms and cell cultures

In Nature

In addition to these biological tools, certain naturally observed configurations of DNA (genes, promoters, enhancers, repressors) and the associated machinery itself are referred to as an expression system. This term is normally used in the case where a gene or set of genes is switched on under well defined conditions. For example the simple repressor switch expression system in Lambda phage and the lac operator system in bacteria. Several natural expression systems are directly used or modified and used for artificial expression systems such as the Tet-on and Tet-off expression system.

Application

The developers of ion semiconductor sequencing have marketed it as a rapid, compact and economical sequencer that can be utilized in a large number of laboratories as a bench top machine. The company hopes that their system will take sequencing outside of specialized centers and into the reach of hospitals and smaller laboratories. A January 2011 New York Times article, "Taking DNA Sequencing to the Masses", underlines these ambitions.

Due to the ability of alternative sequencing methods to achieve a greater read length (and therefore being more suited to whole genome analysis) this technology may be best suited to small scale applications such as microbial genome sequencing, microbial transcriptome sequencing, targeted sequencing, amplicon sequencing, or for quality testing of sequencing libraries.

DNA Nanoball Sequencinge

DNA nanoball sequencing is a type of high throughput sequencing technology used to determine the entire genomic sequence of an organism. The company Complete Genomics uses this technology to sequence samples that researchers submit from several projects. The method uses rolling circle replication to amplify small fragments of genomic

DNA into DNA nanoballs. Unchained sequencing by ligation is then used to determine the nucleotide sequence. This method of DNA sequencing allows large numbers of DNA nanoballs to be sequenced per run and at low reagent costs compared to other next generation sequencing platforms.

However, only short sequences of DNA are determined from each DNA nanoball which makes mapping the short reads to a reference genome difficult.

This technology has been used for multiple genome sequencing projects and is scheduled to be used for more. DNA nanoball sequencing is a high throughput sequencing technology that is used to determine the entire genomic sequence of an organism. The method uses rolling circle replication to amplify small fragments of genomic DNA into DNA nanoballs. Fluorescent probes bind to complementary DNA and the probes are then ligated to anchor sequences bound to known sequences on the DNA template. The base order is determined via the fluorescence of the ligated and bound probes.

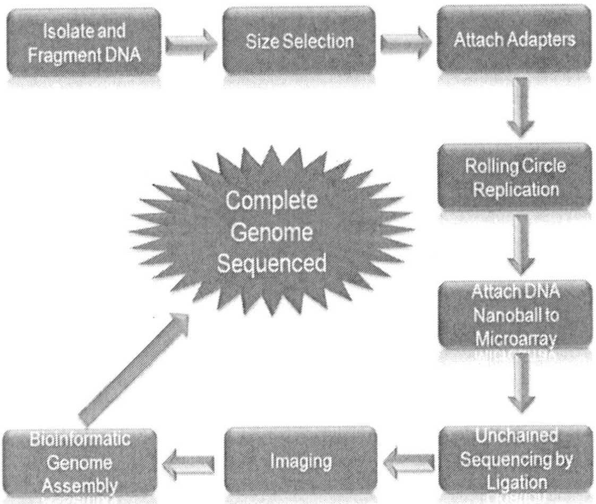

This DNA sequencing method allows large numbers of DNA nanoballs to be sequenced per run at lower reagent costs compared to other next generation sequencing platforms. However, a limitation of this method is that it generates only short sequences of DNA, which presents challenges to mapping its reads to a reference genome. The company Complete Genomics uses DNA nanoball sequencing to sequence samples submitted by researchers.

Procedure

DNA Nanoball Sequencing involves isolating DNA that is to be sequenced, shearing it into small 400 – 500 base pair (bp) fragments, ligating adapter sequences to the fragments, and circularizing the fragments. The circular fragments are copied by rolling circle replication resulting in many single-stranded copies of each fragment. The DNA copies cocatenate head to tail in a long strand, and are compacted into a DNA nanoball. The nanoballs are then adsorbed onto a sequencing flow-cell. Unchained sequencing reactions interrogate specific nucleotide locations in the nanoball by ligating fluorescent probes to the DNA. The color of the fluorescence at each interrogated position is recorded through a high-resolution camera. Bioinformatics are used to analyze the fluorescence data and make a base call, and for mapping the 35-bp mate pair reads to a reference genome. The genome is assembled and any polymorphisms present in the sequence are identified.

DNA Isolation, Fragmentation, and Size Capture

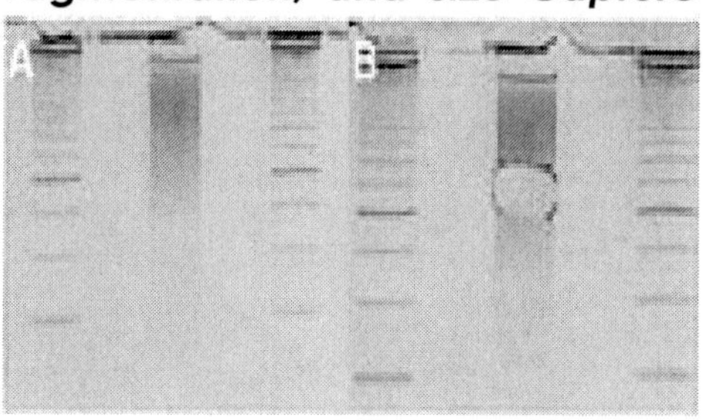

Figure: Polyacrylamide gel electrophoresis (PAGE) gel of a sonicated DNA smear (left) and size selection by gel extraction (right).

Cells are lysed and DNA is extracted from the cell lysate. The high-molecular-weight DNA, often several megabase pairs long, is sonicated to break the DNA double-strands at random intervals. Bioinformatic mapping of the sequencing reads is most efficient when the sample DNA contains a narrow length range. Therefore, selecting the ideal fragment lengths of the DNA for sequencing the fragments are size separated by polyacrylamide gel electrophoresis (PAGE). The DNA of the suitable size range is purified by gel extraction, resulting in DNA with lengths within a narrow range (typically 400 – 500 base pairs).

Attaching Adapter Sequences

Adapter DNA sequences must be attached to the unknown DNA fragment so that DNA with known sequences flank the unknown DNA. In the first round of adapter ligation, a right and left adapter (Ad1) is attached to the right and left flanks of the fragmented DNA, and the DNA is PCR amplified (Figure 3). The right and left Ad1 are modified to create complementary single strand ends that bind to each other and form circular DNA. A restriction enzyme is added which cleaves the DNA 13 bp to the right of the right adapter. This results in linear double-stranded DNA (Figure 3). Right and left adapter sequences (Ad2) are ligated onto the ends of the linear DNA and the product is PCR amplified. The Ad2 sequences are modified to allow them to bind each other and form circular DNA. The restriction enzyme is used again to cleave the circular DNA 13 bp to the left of Ad1. The result is a linear DNA fragment (Figure 3). Right and left adapter sequences (Ad3) are ligated to the right and left flank of the linear DNA and the product is PCR amplified.

The adapters are modified so that they bind to each other and form circular DNA. The type III restriction enzyme EcoP15 is added, which cleaves the DNA 26 bp to the left of Ad3 and 26 bp to the right of Ad2. This step removes a large segment of DNA and linearizes the DNA once again. Right and left adapters (Ad4) are ligated to the DNA, the product is PCR amplified, and the Ad4 sequences are modified so that they bind each other. The result is the completed circular DNA template.

Rolling Circle Replication

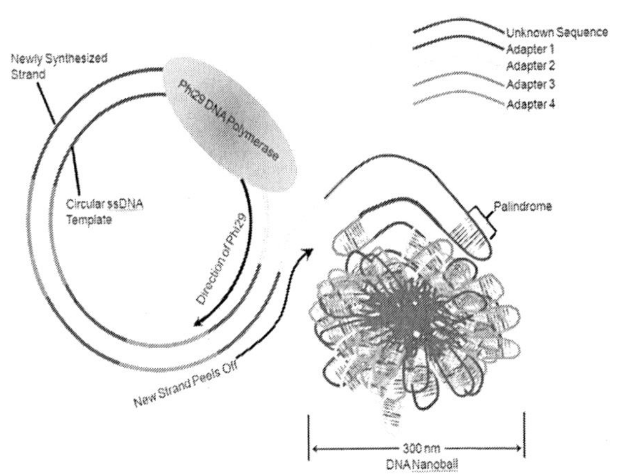

Figure: Rolling Circle Replication To Create DNA Nanoball

Once a circular DNA template, containing sample DNA that is ligated to four unique adapter sequences has been generated, the full sequence is amplified into a long string of DNA. This is accomplished by rolling circle replication with the Phi 29 DNA polymerase which binds and replicates the DNA template (Figure 4). The newly synthesized strand is released from the circular template, resulting in a long single-stranded DNA comprising several head-to-tail copies of the circular template. The four adapter sequences contain palindromic sequences which hybridize and cause the single strand to fold onto itself, resulting in a tight ball of DNA approximately 300 nanometers (nm) across. This allows the nanoballs to remain separated from each other and reduces any tangling between different single stranded DNA lengths.

DNA Nanoball Microarray

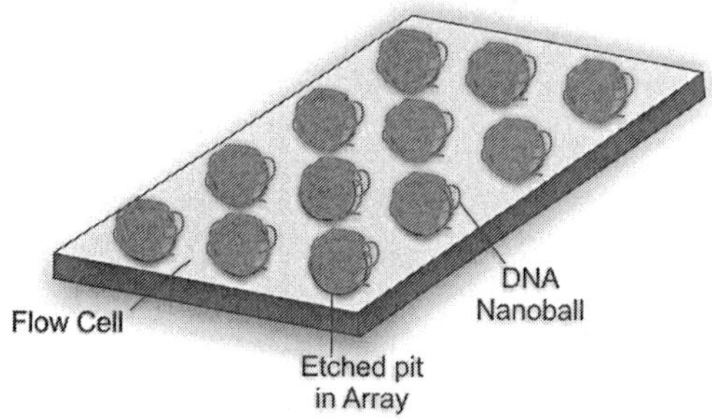

Figure: DNA nanoball Array

To obtain DNA sequence, the DNA nanoballs are attached to a microarray flow cell (Figure 5). The flow cell is a 25 mm by 75 mm silicon wafer coated with silicon dioxide, titanium, hexamethyldisilazane (HMDS), and a photoresist material. The DNA nanoballs are added to the flow cell and selectively bind to the aminosilane in a highly ordered pattern, allowing a very high density of DNA nanoballs to be sequenced.

Unchained sequencing by ligation

The order of the DNA bases between the adapter sequences is determined after being arrayed onto a flow cell. First, oligonucleotide anchor DNA that is complementary to either the right or left end of one of the adapters is added to the flow cell.

Next, T4 DNA ligase is added to a pool of four 10-mer DNA sequences that have degenerate nucleotides in all but one position (for example position 1 next to the anchor, figure) and are added to the flow cell.

The interrogative position in the DNA probe contains an "A" nucleotide with a red fluorophore attached, a "C" with a yellow fluorophore attached, a "G" with a green fluorophore attached or a "T" with a blue fluorophore attached. Only the probe that has a complementary nucleotide in the interrogative position will bind. The T4 DNA ligase attaches the probe to the anchor, the non-binding probes are washed away, and the fluorescence is detected.

The probe/anchor is removed from the DNA nanoball and another anchor is added. A new pool of probes is added with a different interrogative position.

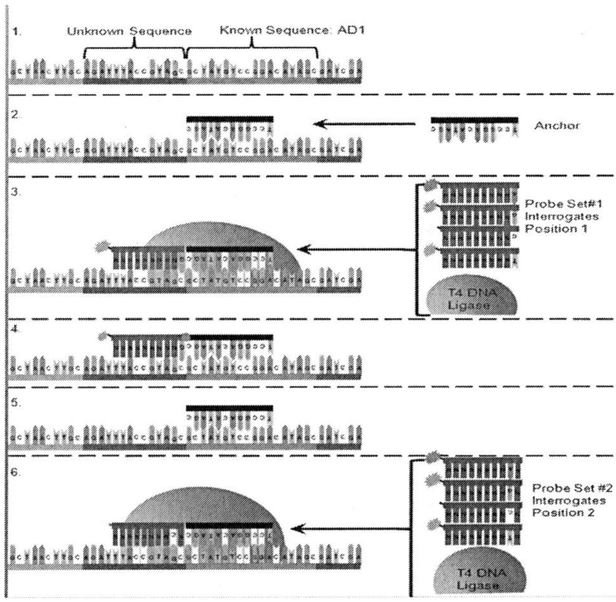

Figure: Unchained Ligation Sequencing

The correct probe hybridizes, is ligated, rinsed and the fluorescence is read and recorded. This process is repeated with all ten interrogation positions next to an anchor sequence.

Once all ten positions are recorded, an anchor is added that binds to a different adapter and the process is repeated to identify the ten nucleotides next to that adapter.

Imaging

After each DNA probe/ligation step, the flow cell is imaged to determine which nucleotide base bound to the DNA nanoball. The fluorophore is excited with an arc lamp that radiates specific wavelengths of light towards the flow cell. The wavelength of the fluorescence of each DNA nanoball is captured on a high resolution CCD camera.

The image is then processed to remove background noise and assess the intensity of each point. The color of each DNA nanoball corresponds to a base at the interrogative position and a computer records the base position information. The order of 70 nucleotides per DNA nanoball is determined using 80 images (8 rounds of 10 interrogative positions per round).

Genome Assembly

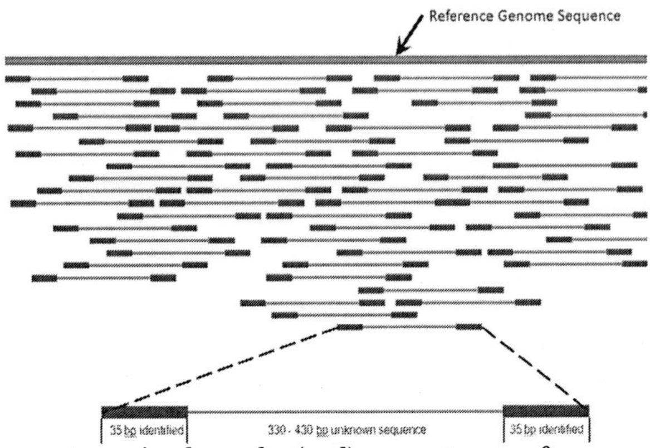

Figure: 35 bp mate paired reads (red) map to a reference genome (dark blue)

In generating the circular template, a large segment of the original 400 – 500 base pair fragment was replaced with the adapter Ad4. The 70 bp that are sequenced are therefore the first 35 bp of the original 400 – 500 bp fragment, and the last 35 bp of the 400 – 500 bp fragment. Therefore, the sequence is identified for two 35 bp reads of DNA separated by about 330 – 430 bp. These 35 bp reads are compared, using bioinformatics, to a reference genome and assigned to a genetic locus (Figure 7). Massive parallel genome mapping is accomplished through high coverage of the reference nucleotide positions, and the complete genome of the DNA sample is assembled. Single base-pair mismatches between the sequenced reads and the reference sequence are used to identify possible single nucleotide polymorphism (SNP). In addition, mapping of one 35-bp portion of the mate pair may identify DNA inserts and deletions (indels) via bioinformatics algorithms that detect possible discrepancies between the mate pairs.

Advantages

DNA nanoball sequencing technology offers several advantages over other sequencing platforms. One major advantage is the use of very high-density arrays. The array design permits one DNA nanoball to attach to each pit that is part of an ordered array, and therefore a higher concentration of DNA can be added. This allows a high percentage of the pits to be occupied by a DNA nanoball thus maximizing the number of reads per flow cell (Figure Top). compared to other sequencing arrays where molecules of DNA are added to a flow cell in a random orientation. Another important advantage is that the sequencing reactions are non-progressive; after each reading of the probe, the probe and anchor are removed and a new anchor and probe set are added. Therefore, if a probe did not bind in the previous reaction, this has no effect on the next probe ligation, thus eliminating a major source of reading error that may occur in other next generation sequencing platforms. It also reduces the use of expensive probes, since DNA nanoball sequencing does not necessitate the probe ligation reaction to be run to completion.

Other advantages of DNA nanoball sequencing include the use of high-fidelity Phi 29 DNA polymerase to ensure accurate amplification of the circular template, several hundred copies of the circular template compacted into a small area resulting in an intense signal, and attachment of the fluorophore to the probe at a long distance from the ligation point results in improved ligation.

Disadvantages

The main disadvantage of DNA nanoball sequencing is short read length of the DNA sequences obtained with this method. Short reads, especially for DNA high in DNA repeats, may map to two or more regions of the reference genome. A second disadvantage of this method is that multiple rounds of PCR have to be used. This can introduce PCR bias and possibly amplify contaminants in the template construction phase.

Applications

DNA nanoball sequencing has been used in recent studies. Lee et al. used this technology to find mutations that were present in a lung cancer and compared them to normal lung tissue. They were able to identify over 50,000 single nucleotide variants Roach et al. used DNA nanoball sequencing to sequence the genomes of a family of four relatives and were able to identify SNPs that may be responsible for a Mendelian disorder, and were able to estimate the inter-generation mutation rate. The Institute for Systems Biology has now uses this technology to sequence 615 complete human genome samples as part of a survery studying neurodegenerative diseases, and the National Cancer Institute is using DNA nanoball sequencing to sequence 50 tumours and matched normal tissues from pediatric cancers.

Significance

Massively parallel next generation sequencing platforms like DNA nanoball sequencing may contribute to the treatment and diagnosis of many genetic diseases. The cost of sequencing an entire human genome has fallen from about one million dollars in 2008, to $4400 dollars in 2010 with the DNA nanoball technology. Sequencing of the entire genome of patients with heritable diseases or cancer, mutations associated with these diseases have been identified, opening up strategies, such as targeted therapeutics for at-risk people and for genetic counseling. As the price of sequencing an entire human genome approaches the $1000 mark, genomic sequencing the entire population may become feasible as part of normal preventative medicine.

Helioscope(TM) Single Molecule Sequencing

Based on "true single molecule sequencing" technology, Helioscope sequencing uses DNA fragments with added polyA tail adapters, which are attached to the flow cell surface. The next steps involve extension-based sequencing with cyclic washes of the flow cell with fluorescently labeled nucleotides (one nucleotide type at a time, as with the Sanger method). The reads are performed by the Helioscope sequencer. The reads are short, up to 55 bases per run, but recent improvemend of the methodology allowes more accurate reads of homopolymers (stretches of one type of nucleotides) and RNA sequencing.

Single Molecule SMRT(TM) Sequencing

SMRT sequencing is based on the sequencing by synthesis approach. The DNA is synthesisd in so called zero-mode wave-guides (ZMWs) - small well-like containers with the capturing tools located at the bottom of the well.

The sequencing is performed with use of unmodified polymerase (attached to the ZMW bottom) and fluorescently labelled nucleotides flowing freely in the solution. The wells are constructed in a way that only the fluorescence occurring by the bottom of the well is detected. The fluorescent label is detached from the nucleotide at its incorporation into the DNA strand, leaving an unmodified DNA strand. According to Pacific Biosciences, the SMTR technology developer, this methodology allows detection of nucleotide modifications (such as cytosine methylation). This happens through the observation of polymerase kinetics. This approach allows reads of 1000 nucleotides.

Single Molecule Real Time (RNAP) Sequencing

This method is based on RNA polymerase (RNAP), which is attached to a polystyrene bead, with distal end of sequenced DNA is attached to another bead, with both beads being placed in optical traps. RNAP motion during transcription brings the beads in closer and their relative distance changes, which can then be recorded at a single nucleotide resolution. The sequence is deduced based on the four readouts with lowered concentrations of each of the four nucleotide types (similarly to Sangers method).

Nanopore DNA Sequencing

This method is based on the readout of electrical signal occurring at nucleotides passing by alpha-hemolysin pores covalently bound with cyclodextrin. The DNA passing through the nanopore changes its ion current. This change is dependent on the shape, size and length of the DNA sequence. Each type of the nucleotide blocks the ion flow through the pore for a different period of time. The method has a potential of development as it does not require modified nucleotides, however single nucleotide resolution is not yet available.

VisiGen Biotechnologies Approach

VisiGen Biotechnologies introduced a specially engineered DNA polymerase for use in their sequencing. This polymerase acts as a sensor - having incorporated a donor fluorescent dye by its active centre. This donor dye acts by FRET (fluorescent resonant energy transfer), inducing fluorescence of differently labeled nucleotides. This approach allows reads performed at the spead at which polymerase incorporates nucleotides into the sequence (several hundred per second). The nucleotide fluorochrome is released after the incorporation into the DNA strand. The expected read lengths in this approach should reach 1000 nucleotides, however this will have to be confirmed.

Future Methods

Sequencing by hybridization is a non-enzymatic method that uses a DNA microarray. A single pool of DNA whose sequence is to be determined is fluorescently labeled and hybridized to an array containing known sequences. Strong hybridization signals from a given spot on the array identifies its sequence in the DNA being sequenced. Mass spectrometry may be used to determine mass differences between DNA fragments produced in chain-termination reactions.

DNA sequencing methods currently under development include labeling the DNA polymerase, reading the sequence as a DNA strand transits through nanopores, and microscopy-based techniques, such as AFM or transmission electron microscopy that are used to identify the positions of individual nucleotides within long DNA fragments (>5,000 bp) by nucleotide labeling with heavier elements (e.g., halogens) for visual detection and recording. Third generation technologies aim to increase throughput and decrease the time to result and cost by eliminating the need for excessive reagents and harnessing the processivity of DNA polymerase. In microfluidic Sanger sequencing the entire thermocycling amplification of DNA fragments as well as their separation by electrophoresis is done on a single glass wafer (approximately 10 cm in diameter) thus reducing the reagent usage as well as cost. In some instances researchers have shown that they can increase the throughput of conventional sequencing through the use of microchips. Research will still need to be done in order to make this use of technology effective.

In October 2006, the X Prize Foundation established an initiative to promote the development of full genome sequencing technologies, called the Archon X Prize, intending to award $10 million to "the first Team that can build a device and use it to sequence 100 human genomes within 10 days or less, with an accuracy of no more than one error in every 100,000 bases sequenced, with sequences accurately covering at least 98% of the genome, and at a recurring cost of no more than $10,000 (US) per genome."

Each year NHGRI promotes grants for new research and developments in genomics. 2010 grants and 2011 candidates include continuing work in microfluidic, polony and base-heavy sequencing methodologies

Major Landmarks in DNA Sequencing

- 1953 Discovery of the structure of the DNA double helix.
- 1972 Development of recombinant DNA technology, which permits isolation of defined fragments of DNA; prior to this, the only accessible samples for sequencing were from bacteriophage or virus DNA.
- 1977 The first complete DNA genome to be sequenced is that of bacteriophage öX174.
- 1977 Allan Maxam and Walter Gilbert publish "DNA sequencing by chemical degradation". Frederick Sanger, independently, publishes "DNA sequencing with chain-terminating inhibitors".
- 1984 Medical Research Council scientists decipher the complete DNA sequence of the Epstein-Barr virus, 170 kb.
- 1986 Leroy E. Hood's laboratory at the California Institute of Technology and Smith announce the first semi-automated DNA sequencing machine.
- 1987 Applied Biosystems markets first automated sequencing machine, the model ABI 370.

- 1990 The U.S. National Institutes of Health (NIH) begins large-scale sequencing trials on Mycoplasma capricolum, Escherichia coli, Caenorhabditis elegans, and Saccharomyces cerevisiae (at US$0.75/base).
- 1991 Sequencing of human expressed sequence tags begins in Craig Venter's lab, an attempt to capture the coding fraction of the human genome.
- 1995 Craig Venter, Hamilton Smith, and colleagues at The Institute for Genomic Research (TIGR) publish the first complete genome of a free-living organism, the bacterium Haemophilus influenzae. The circular chromosome contains 1,830,137 bases and its publication in the journal Science marks the first use of whole-genome shotgun sequencing, eliminating the need for initial mapping efforts.
- 1996 Pål Nyrén and his student Mostafa Ronaghi at the Royal Institute of Technology in Stockholm publish their method of pyrosequencing
- 1998 Phil Green and Brent Ewing of the University of Washington publish "phred" for sequencer data analysis.
- 2000 Lynx Therapeutics publishes and markets "MPSS" - a parallelized, adapter/ligation-mediated, bead-based sequencing technology, launching "next-generation" sequencing.
- 2001 A draft sequence of the human genome is published.
- 2004 454 Life Sciences markets a parallelized version of pyrosequencing. The first version of their machine reduced sequencing costs 6-fold compared to automated Sanger sequencing, and was the second of a new generation of sequencing technologies, after MPSS.
- Genetic disorder

A genetic disorder is an illness caused by abnormalities in genes or chromosomes, especially a condition that is present from before birth. Most genetic disorders are quite rare and affect one person in every several thousands or millions.

A genetic disorder may or may not be a heritable disorder. Some genetic disorders are passed down from the parents' genes, but others are always or almost always caused by new mutations or changes to the DNA. In other cases, the same disease, such as some forms of cancer, may be caused by an inherited genetic condition in some people, by new mutations in other people, and by non-genetic causes in still other people.

Some types of recessive gene disorders confer an advantage in certain environments when only one copy of the gene is present.

Single Gene Disorder

Table: Disorder Prevalence (approximate)

Disorder	Prevalence
Autosomal dominant	
Familial hypercholesterolemia	1 in 500
Polycystic kidney disease	1 in 1250
Neurofibromatosis Type I	1 in 2,500
Hereditary spherocytosis	1 in 5,000
Marfan syndrome	1 in 4,000
Huntington's disease 1 in 15,000	
Autosomal recessive	
Sickle cell anemia	1 in 625
(African) Cystic fibrosis 1 in 2,000	
(Caucasians)	
Lysosomal Acid Lipase (LAL) Deficiency	1 in 40,000
Tay-Sachs disease	1 in 3,000
(American Jews)	
Phenylketonuria	1 in 12,000
Mucopolysaccharidoses	1 in 25,000
Glycogen storage diseases	1 in 50,000
Galactosemia	1 in 57,000
X-linked	
Duchenne muscular dystrophy	1 in 7,000
Hemophilia	1 in 10,000

Values are for liveborn infants

A single gene disorder is the result of a single mutated gene. There are estimated to be over 4000 human diseases caused by single gene defects. Single gene disorders can be passed on to subsequent generations in several ways. Genomic imprinting and uniparental disomy, however, may affect inheritance patterns. The divisions between recessive and dominant types are not "hard and fast" although the divisions between autosomal and X-linked types are (since the latter types are distinguished purely based on the chromosomal location of the gene). For example, achondroplasia is typically considered a dominant disorder, but children with two genes for achondroplasia have a severe skeletal disorder that achondroplasics could be viewed as carriers of. Sickle-cell anemia is also considered a recessive condition, but heterozygous carriers have increased resistance to malaria in early childhood, which could be described as a related dominant condition. When a couple where one partner or both are sufferers or carriers of a single gene disorder and wish to have a child they can do so through IVF which means they can then have PGD (pre-implantation genetic diagnosis) to check whether the fertilised egg has had the genetic disorder passed on.

Autosomal Dominant

Only one mutated copy of the gene will be necessary for a person to be affected by an autosomal dominant disorder. Each affected person usually has one affected parent. There is a 50% chance that a child will inherit the mutated gene. Conditions that are autosomal dominant sometimes have reduced penetrance, which means that although only one mutated copy is needed, not all individuals who inherit that mutation go on to develop the disease. Examples of this type of disorder are Huntington's disease, neurofibromatosis type 1, neurofibromatosis type 2, Marfan syndrome, hereditary nonpolyposis colorectal cancer, and hereditary multiple exostoses, which is a highly penetrant autosomal dominant disorder. Birth defects are also called congenital anomalies.

Dominance (genetics)

Dominance in genetics is a relationship between two variant forms (alleles) of a single gene, in which one allele masks the effect of the other in influencing some trait. In the simplest case, if a gene exists in two allelic forms (A & a), three combinations of alleles (genotypes) are possible: AA, Aa, and aa. If Aa individuals (heterozygotes) show the same form of the trait (phenotype) as AA individuals (homozygotes), and aa homozygotes show a different phenotype, allele A is said to dominate or be dominant to allele a, and a is said to be recessive to A. By one convention, dominant alleles are written in uppercase letters, and recessive alleles in lowercase letters.

Background

Diploid / Haploid

Most familiar plants, like peas, and familiar animals, like fruit flies and humans, have paired chromosomes, and are described as diploid. One chromosome of each pair is contributed by each parent: one by the female parent in her ovum, and one by the male parent in his sperm, which are joined at fertilization. The ovum and sperm cells have only one copy of each chromosome and are described as haploid. Production of haploid gametes occurs through a process called meiosis.

Chromosomes, Genes, and Alleles

Each chromosome of a matching pair is structurally similar to the other, and each member of a homologous pair has the same genetic material arranged in the same order and physical locations (loci, singular locus). The genetic material in each chromosome comprises a series of discrete genes that influence various traits. Thus, each gene also has a corresponding homologue, which may exist in different forms: the variant forms are called alleles. The alleles at the same locus on the two homologous chromosomes may be identical or different.

In popular usage, "gene" and "allele" are often used interchangeably. This produces misunderstandings. Properly, "gene" refers to a hereditary unit, ordinarily at a fixed position on a chromosome, that influences a particular trait. Genes are now understood to comprise DNA. "Allele" refers to any of the many particular forms of a gene that may be present in a population of individuals from a particular species, at a particular locus. E.g., it is inaccurate to say "This pea plant has a pair of wrinkled genes", and it is more accurate to say, "This plant has two 'w' alleles for the 'Seed Shape' gene, and will produce wrinkled peas." Consider also the example of blood type in humans. Near the long arm of chromosome nine appears a gene that determines whether an individual will be blood type, A, B, or O. There are three different alleles that could be present at this locus, but only two can be present in any individual, one inherited from their mother and one from their father.

Homozygous, Heterozygous

If two alleles of a given gene are identical, the organism is called a homozygote and is homozygous with respect to that gene; if instead the two alleles are different, the organism is a heterozygote and is heterozygous. The genetic makeup of an organism, either at a single locus or over all its genes collectively, is called the genotype. The genotype of an organism directly or indirectly affects its molecular, physical,and other traits, which individually or collectively are called the phenotype. At heterozygous gene loci, the two alleles interact to produce the phenotype. The simplest form of allele interaction is the one described by Mendel, now called Mendelian, in which the appearance/phenotype caused by one allele is apparent, called dominant, and the appearance/phenotype caused by the other allele is not apparent, called recessive.

In the simplest case, the phenotypic effect of one allele completely masks the other in heterozygous combination; that is, the phenotype produced by the two alleles in heterozygous combination is identical to that produced by one of the two homozygous genotypes. The allele that masks the other is said to be dominant to the latter, and the alternative allele is said to be recessive to the former.

Which Trait is Dominant?

The terms dominant and recessive refer to the interaction of alleles in producing the phenotype of the heterozygote. If there are two alternative phenotypes, by definition the phenotype exhibited by the heterozygote is called "dominant" and the "hidden" phenotype is called "recessive". The key concept of dominance is that the heterozygote is phenotypically identical to one of the two homozygotes. That trait corresponding to the dominant allele may then be called the "dominant" trait.

Dominance is a genotypic relationship between alleles, as manifested in the phenotype. It is unrelated to the nature of the phenotype itself, e.g., whether it is regarded as normal or abnormal, standard or nonstandard, healthy or diseased, stronger or weaker, or more or less extreme. It is also important to distinguish between the "round" gene locus, the "round" allele at that locus, and the "round" phenotype it produces. It is inaccurate to say that "the round gene dominates the wrinkled gene" or that "round peas dominate wrinkled peas."

Nomenclature

In genetics, the common convention is that dominant alleles are written as capital letters and recessive alleles as lower-case letters. In the pea example, once the dominance relationships of the two alleles are known, it is possible to designate the dominant allele that produces a round shape by a capital-letter symbol R, and the alternative recessive allele that produces a wrinkled shape by a lower-case symbol r. The homozygous dominant, heterozygous, and homozygous recessive genotypes are then written RR, Rr, and rr, respectively. It would also be possible to designate the two alleles as W and w, and the three genotypes WW, Ww, and ww, the first two of which produced round peas and the third wrinkled peas. Note that the choice or "R" or "W" as the symbol for the dominant allele does not pre-judge whether the allele causing the "round" or "wrinkled" phenotype when homozygous is the dominant one.

Another system of notation designates the gene involved in seed shape as the "Shp" gene, which exists in two allelic forms, Shp and Shp , the dominance relationships of the two being indicated by the case of the superscripts. This system is the standard system in Drosophila genetics.

Relationship To Other Genetic Concepts

The concept of dominance is involved with a number of other genetic concepts.

Multiple Alleles

Although any individual has at most two different alleles, most genes exist in a large number of allelic forms in the population as a whole. In some cases, the alleles have different effects on the phenotype, and their dominance interactions with each other can be described as a series. For example, the best known human blood groups, the ABO system, comprises three sets of alleles at the I locus, I , I , and I . The first two are dominant to the latter: that is, the AA and AO genotypes produce indistinguishable blood group phenotypes, called "Type A", as do BB and BO, which produce "Type B" blood. In another example, coat color in siamese cats and related breeds is determined by a series of alleles at the albino gene locus (c) that produce different levels of pigment and hence different levels of color dilution. Four of these are c , c , c , and c (standard, Burmese, siamese, and albino, respectively), where the first allele is completely dominant to the last three, and the last is completely recessive to the first three.

Incomplete and Semi-dominance

Complete dominance occurs when the phenotype of the heterozygote is completely indistinguishable from that of the dominant homozygote. This is frequently not the case. Incomplete dominance occurs when the phenotype of the heterozygous genotype is an intermediate of the phenotypes of the homozygous genotypes. For example, the snapdragon flower color is either homozygous for red or white. When the red homozygous flower is paired with the white homozygous flower, the result yields a pink snapdragon flower. The pink snapdragon is the result of incomplete dominance.

Co-dominance

Codominance refers to a relationship between two alleles of a gene. It occurs when the contributions of both alleles at a single locus are visible and do not overpower each other in the phenotype.

For instance, in the ABO system, the I and I alleles are co-dominant in producing the AB blood group phenotype, in which both A- and B-type antigens are made. Another example occurs at the locus for the Beta-globin component of hemoglobin, where the three molecular phenotypes of Hb /Hb , Hb /Hb , and Hb /Hb are all equally detectable by protein electrophoresis. (The medical condition produced by the heterozygous genotype is called an incomplete dominant). For most gene loci at the molecular level, both alleles are expressed co-dominantly, because both are transcribed into RNA.

Co-dominance and incomplete or semi-dominance are not the same thing. For example, in some plant species, pink flowers may be the product of a mixture of red and white pigments (co-dominance on the pigment level, no dominance on the color level), or the result of one allele that produces the usual amount of red pigment and another non-functional allele that produces no pigment, so as to produce a dilute, intermediate pink color (no dominance at either level).

Blood Types in Humans Show Co-dominance

Co-dominance occurs when the contributions of both alleles are visible in the phenotype. In the ABO example, the I and I alleles are co-dominant in producing the AB blood group phenotype, in which both A-type and B-type antigens are made. Another example occurs at the locus for the Beta-globin component of hemoglobin, where the three molecular phenotypes of Hb /Hb , Hb /Hb , and Hb /Hb are all equally detectable by protein electrophoresis. (The medical condition produced by the heterozygous genotype is called an incomplete dominant. For most gene loci at the molecular level, both alleles are expressed co-dominantly, because both are transcribed into RNA.

Co-dominance is different from incomplete or semi-dominance. For example, pink flowers might be the product of two alleles that produce red and white pigments that become mixed (co-dominance on the pigment level, no dominance on the color level), or the result of one allele that produces the usual amount of red pigment and another non-functional allele that produces no pigment, so as to produce a dilute, intermediate pink color (no dominance at either level).

Autosomal Versus Sex-linked Dominance

In humans and other mammal species, sex is determined by two sex chromosomes called the X chromosome and the Y chromosome. Human females are typically XX; males are typically XY. The remaining pairs of chromosome are found in both sexes and are called autosomes; genetic traits due to loci on these chromosomes are described as autosomal, and may be dominant or recessive. Genetic traits on the X and Y chromosomes are called sex linked, because they tend to be characteristic of one sex or the other. In practice, the term almost always refers to X-linked traits. Females have two copies of every gene locus found on the X chromosome, just as for the autosomes, and the same dominance relationships apply. Males however have only one copy of each X chromosome gene locus, and are described as hemizygous for these genes. The Y chromosome is much smaller than the X, and contains a much smaller set of genes that influence 'maleness', such as the SRY gene for testis determining factor. Dominance rules for sex-linked gene loci are determined by their behavior in the female: because the male has only one allele, that allele is always expressed regardless of whether it is dominant or recessive. An example of sex-linked is: Hair loss is a trait caused by a gene located on the X chromosomes.

Epistasis

Epistasis ["epi + stasis = to sit on top"] is an interaction between genotypes at two different gene loci, which sometimes resembles a dominance interaction at a single locus.

Epistasis modifies the characteristic 9:3:3:1 ratio expected for two non-epistatic genes. Most genetic systems involve complex epistatic interactions among multiple gene loci. For two loci, 14 classes of epistatic interactions are recognized. As an example of recessive epistasis, one gene locus may determine whether a flower pigment is yellow (AA or Aa) or green (aa), while another locus determines whether the pigment is produced (BB or Bb) or not (bb). In a bb plant, the flowers will be white, irrespective of the genotype of the other locus as AA, Aa, or aa.

The b allele is not dominant to the A allele: the B locus shows recessive epistasis to the A locus, because the B locus when homozygous for the recessive allele (bb) suppresses phenotypic expression of the A locus. In a cross between two AaBb plants, this produces a characteristic 9:3:4 ratio, in this case of yellow : green : white flowers. In dominant epistasis, one gene locus may determine yellow and green pigment as in the previous example: AA and Aa are yellow, and aa are green. A second locus determines whether a pigment precursor is produced (dd) or not (DD or Dd). Here, in a D- plant, the flowers will be colorless irrespective of the genotype at the A locus, because of the epistatic effect of the dominant D allele. Thus, in a cross between two AaDd plants, 3/4 of the plants will be colorless, and the yellow and green phenotypes are expressed only in dd plants. This produces a characteristic 12:3:1 ratio of white : yellow : green plants.

Supplementary epistasis occurs when two loci affect the same phenotype. For example, if pigment color is produced by CC or Cc but not cc, and by DD or Dd but not dd, then pigment is produced only in C-D- genotypes, and not in any genotype combination with cc or dd. That is, both loci must have at least one dominant allele to produce the phenotype. This produces a characteristic ratio 9:7 ratio of pigmented to unpigmented plants. Classification by fitness or trait value

Diagram illustrating different relationships between numbers of mutations and fitness. Synergistic epistasis is the blue line - each mutation has a disproportionately large effect on the organism's fitness. Antagonistic epistasis is the red line.

See Evolution of Sex Two-locus epistatic interactions can be either synergistic (enhancing the effectiveness) or antagonistic (reducing the activity).

In the example of a haploid organism with genotypes (at two loci) AB, Ab, aB or ab, we can think of the following trait values where higher values suggest greater expression of the characteristic (the exact values are simply given as examples):

	AB	**Ab**	**aB**	**ab**
No epistasis (additive across loci)	*2*	*1*	*1*	*0*
Synergistic epistasis	*3*	*1*	*1*	*0*
Antagonistic epistasis	*1*	*1*	*1*	*0*

Hence, we can classify thus:

Trait Values	**Type of Epistasis**
AB = Ab + aB " ab	*No epistasis, additive inheritance*
AB > Ab + aB *e* ab	*Synergistic epistasis*
AB < Ab + aB *e* ab	*Antagonistic epistasis*

Understanding whether the majority of genetic interactions are synergistic or antagonistic will help solve such problems as the evolution of sex.

Epistasis and Sex

Negative epistasis and sex are thought to be intimately correlated. Experimentally, this idea has been tested in using digital simulations of asexual and sexual populations.

Over time, sexual populations move towards more negative epistasis, or the lowering of fitness by two interacting alleles. It is thought that negative epistasis allows individuals carrying the interacting deleterious mutations to be removed from the populations efficiently. This removes those alleles from the population, resulting in an overall more fit population. This hypothesis was proposed by Alexey Kondrashov, and is sometimes known as the deterministic mutation hypothesis and has also been tested using artificial gene networks. However, the evidence for this hypothesis has not always been straightforward and the model proposed by Kondrashov has been criticized for assuming mutation parameters far from real world observations.

For example, MacCarthy and Bergman. In addition, in those tests which used artificial gene networks, negative epistasis is only found in more densely connected networks, whereas empirical evidence indicates that natural gene networks are sparsely connected, and theory shows that selection for robustness will favor more sparsely connected and minimally complex networks.

5

Elements of Chemogenomics

Genetic testing (also called DNA-based tests) is among the newest and most sophisticated of techniques used to test for genetic disorders which involves direct examination of the DNA molecule itself. Other genetic tests include biochemical tests for such gene products as enzymes and other proteins and for microscopic examination of stained or fluorescent chromosomes. Genetic tests are used for several reasons, including:

- Carrier screening, which involves identifying unaffected individuals who carry one copy of a gene for a disease that requires two copies for the disease to be expressed
- Preimplantation genetic diagnosis.
- Prenatal diagnostic testing
- Newborn screening
- Genealogical DNA test (for genetic genealogy purposes)
- Presymptomatic testing for predicting adult-onset disorders such as Huntington's disease
- Presymptomatic testing for estimating the risk of developing adult-onset cancers and Alzheimer's disease
- Confirmational diagnosis of a symptomatic individual
- Forensic/identity testing

Genetic testing allows the genetic diagnosis of vulnerabilities to inherited diseases, and can also be used to determine a child's paternity (genetic father) or a person's ancestry. Normally, every person carries two copies of every gene (with the exception of genes related to sex-linked traits, which are only inherited from the mother by males), one inherited from their mother, one inherited from their father. The human genome is believed to contain around 20,000 - 25,000 genes. In addition to studying chromosomes to the level of individual genes, genetic testing in a broader sense includes biochemical tests for the possible presence of genetic diseases, or mutant forms of genes associated with increased risk of developing genetic disorders. Genetic testing identifies changes in chromosomes, genes, or proteins. Most of the time, testing is used to find changes that are associated with inherited disorders.

The results of a genetic test can confirm or rule out a suspected genetic condition or help determine a person's chance of developing or passing on a genetic disorder. Several hundred genetic tests are currently in use, and more are being developed.

Since genetic testing may open up ethical or psychological problems, genetic testing is often accompanied by genetic counseling.

Types

Genetic testing is "the analysis of, chromosomes (DNA), proteins, and certain metabolites in order to detect heritable disease-related genotypes, mutations, phenotypes, or karyotypes for clinical purposes." It can provide information about a person's genes and chromosomes throughout life. Available types of testing include:

- Newborn screening: Newborn screening is used just after birth to identify genetic disorders that can be treated early in life. The routine testing of infants for certain disorders is the most widespread use of genetic testing—millions of babies are tested each year in the United States. All states currently test infants for phenylketonuria (a genetic disorder that causes mental illness if left untreated) and congenital hypothyroidism (a disorder of the thyroid gland).
- Diagnostic testing: Diagnostic testing is used to diagnose or rule out a specific genetic or chromosomal condition. In many cases, genetic testing is used to confirm a diagnosis when a particular condition is suspected based on physical mutations and symptoms. Diagnostic testing can be performed at any time during a person's life, but is not available for all genes or all genetic conditions. The results of a diagnostic test can influence a person's choices about health care and the management of the disease.
- Carrier testing: Carrier testing is used to identify people who carry one copy of a gene mutation that, when present in two copies, causes a genetic disorder. This type of testing is offered to individuals who have a family history of a genetic disorder and to people in ethnic groups with an increased risk of specific

genetic conditions. If both parents are tested, the test can provide information about a couple's risk of having a child with a genetic condition.

- Prenatal testing: Prenatal testing is used to detect changes in a fetus's genes or chromosomes before birth. This type of testing is offered to couples with an increased risk of having a baby with a genetic or chromosomal disorder. In some cases, prenatal testing can lessen a couple's uncertainty or help them decide whether to abort the pregnancy. It cannot identify all possible inherited disorders and birth defects, however.
- Preimplantation genetic diagnosis: Genetic testing procedures that are performed on human embryos prior to the implantation as part of an in vitro fertilization procedure.
- Predictive and presymptomatic testing: Predictive and presymptomatic types of testing are used to detect gene mutations associated with disorders that appear after birth, often later in life. These tests can be helpful to people who have a family member with a genetic disorder, but who have no features of the disorder themselves at the time of testing. Predictive testing can identify mutations that increase a person's chances of developing disorders with a genetic basis, such as certain types of cancer. For example, an individual with a mutation in BRCA1 has a 65% cumulative risk of breast cancer. Presymptomatic testing can determine whether a person will develop a genetic disorder, such as hemochromatosis (an iron overload disorder), before any signs or symptoms appear. The results of predictive and presymptomatic testing can provide information about a person's risk of developing a specific disorder and help with making decisions about medical care.
- Forensic testing: Forensic testing uses DNA sequences to identify an individual for legal purposes. Unlike the tests described above, forensic testing is not used to detect gene mutations associated with disease. This type of testing can identify crime or catastrophe victims, rule out or implicate a crime suspect, or establish biological relationships between people (for example, paternity).
- Parental testing: This type of genetic test uses special DNA markers to identify the same or similar inheritance patterns between related individuals. Based on the fact that we all inherit half of our DNA from the father, and half from the mother, DNA

scientists test individuals to find the match of DNA sequences at some highly differential markers to draw the conclusion of relatedness.

- Research testing: Research testing includes finding unknown genes, learning how genes work and advancing our understanding of genetic conditions. The results of testing done as part of a research study are usually not available to patients or their healthcare providers.
- Pharmacogenomics: type of genetic testing that determines the influence of genetic variation on drug response.

Medical Procedure

Genetic testing is often done as part of a genetic consultation and as of mid-2008 there were more than 1,200 clinically applicable genetic tests available. Once a person decides to proceed with genetic testing, a medical geneticist, genetic counselor, primary care doctor, or specialist can order the test after obtaining informed consent.

Genetic tests are performed on a sample of blood, hair, skin, amniotic fluid (the fluid that surrounds a fetus during pregnancy), or other tissue. For example, a medical procedure called a buccal smear uses a small brush or cotton swab to collect a sample of cells from the inside surface of the cheek. Alternatively, a small amount of saline mouthwash may be swished in the mouth to collect the cells. The sample is sent to a laboratory where technicians look for specific changes in chromosomes, DNA, or proteins, depending on the suspected disorder. The laboratory reports the test results in writing to a person's doctor or genetic counselor.

Routine newborn screening tests are done on a small blood sample obtained by pricking the baby's heel with a lancet.

Interpreting Results

The results of genetic tests are not always straightforward, which often makes them challenging to interpret and explain. When interpreting test results, healthcare professionals consider a person's medical history, family history, and the type of genetic test that was done.

A positive test result means that the laboratory found a change in a particular gene, chromosome, or protein of interest. Depending on the purpose of the test, this result may confirm a diagnosis, indicate that a person is a carrier of a particular genetic mutation, identify an increased risk of developing a disease (such as cancer) in the future, or suggest a need for further testing. Because family members have some genetic material in common, a positive test result may also have implications for certain blood relatives of the person undergoing testing. It is important to note that a positive result of a predictive or presymptomatic genetic test usually cannot establish the exact risk of developing a disorder. Also, health professionals typically cannot use a positive test result to predict the course or severity of a condition.

A negative test result means that the laboratory did not find a dangerous copy of the gene, chromosome, or protein under consideration. This result can indicate that a person is not affected by a particular disorder, is not a carrier of a specific genetic mutation, or does not have an increased risk of developing a certain disease. It is possible, however, that the test missed a disease-causing genetic alteration because many tests cannot detect all genetic changes that can cause a particular disorder. Further testing may be required to confirm a negative result.

In some cases, a negative result might not give any useful information. This type of result is called uninformative, indeterminate, inconclusive, or ambiguous. Uninformative test results sometimes occur because everyone has common, natural variations in their DNA, called polymorphisms, that do not affect health. If a genetic test finds a change in DNA that has not been associated with a disorder in other people, it can be difficult to tell whether it is a natural polymorphism or a disease-causing mutation. An uninformative result cannot confirm or rule out a specific diagnosis, and it cannot indicate whether a person has an increased risk of developing a disorder. In some cases, testing other affected and unaffected family members can help clarify this type of result.

Risks and Limitations

The physical risks associated with most genetic tests are very small, particularly for those tests that require only a blood sample or buccal smear (a procedure that samples cells from the inside surface of the cheek). The procedures used for prenatal testing carry a small but real risk of losing the pregnancy (miscarriage) because they require a sample of amniotic fluid or tissue from around the fetus. Many of the risks associated with genetic testing involve the emotional, social, or financial consequences of the test results. People may feel angry, depressed, anxious, or guilty about their results. In some cases, genetic testing creates tension within a family because the results can reveal information about other family members in addition to the person who is tested. The possibility of genetic discrimination in employment or insurance is also a concern. Some individuals avoid genetic testing out of fear it will affect their ability to purchase insurance or find a job.

Health insurers do not currently require applicants for coverage to undergo genetic testing, and when insurers encounter genetic information, it is subject to the same confidentiality protections as any other sensitive health information. In the United States, the use of genetic information is governed by the Genetic Information Nondiscrimination Act (GINA).

Genetic testing can provide only limited information about an inherited condition. The test often can't determine if a person will show symptoms of a disorder, how severe the symptoms will be, or whether the disorder will progress over time. Another major limitation is the lack of treatment strategies for many genetic disorders once they are diagnosed. A genetics professional can explain in detail the benefits, risks, and limitations of a particular test. It is important that any person who is considering genetic testing understand and weigh these factors before making a decision.

Direct-to-Consumer Genetic Testing

Direct-to-Consumer (DTC) genetic testing is a type of genetic test that is accessible directly to the consumer without having to go through a health care professional. Usually, to obtain a genetic test, health care professionals such as doctors acquire the permission of the patient and order the desired test.

DTC genetic tests, however, allow consumers to bypass this process and order one themselves. There are a variety of DTC tests, ranging from testing for breast cancer alleles to mutations linked to cystic fibrosis. Benefits of DTC testing are the accessibility of tests to consumers, promotion of proactive healthcare and the privacy of genetic information. Possible additional risks of DTC testing are the lack of governmental regulation and the potential misinterpretation of genetic information.

Controversy

DTC genetic testing has been controversial due to outspoken opposition within the scientific community. Critics of DTC testing argue against the risks involved, the unregulated advertising and marketing claims, and the overall lack of governmental oversight. DTC testing involves many of the same risks associated with any genetic test. One of the more obvious and dangerous of these is the possibility of severe misreading of test results. Without professional guidance, consumers can potentially misinterpret genetic information, causing them to be deluded about their personal health.

Some advertising for direct-to-consumer genetic testing has been criticized as conveying an exaggerated and inaccurate message about the connection between genetic information and disease risk, utilizing emotions as a selling factor. An advertisement for a BRCA-predictive genetic test for breast cancer stated: "There is no stronger antidote for fear than information."

Government Regulation in the United States

Currently, the U.S. has no strong Federal regulation moderating the DTC market. Though there are several hundred tests available, only a handful are approved by the Food and Drug Administration (FDA); these are sold as at-home test kits, and are therefore considered "medical devices" over which the FDA may assert jurisdiction.

Other types of DTC tests require customers to mail in DNA samples for testing; it is difficult for the FDA to exercise jurisdiction over these types of tests, because the actual testing is completed in the laboratories of providers. As of 2007, the FDA had not yet officially substantiated with scientific evidence the claimed accuracy of the majority of direct-to-consumer genetic tests.

With regard to genetic testing and information in general, legislation in the United States called the Genetic Information Nondiscrimination Act prohibits group health plans and health insurers from denying coverage to a healthy individual or charging that person higher premiums based solely on a genetic predisposition to developing a disease in the future.

The legislation also bars employers from using individuals' genetic information when making hiring, firing, job placement, or promotion decisions. The legislation, the first of its kind in the U.S., was passed by the United States Senate on April 24, 2008, on a vote of 95-0, and was signed into law by President George W. Bush on May 21, 2008. It went into effect on November 21, 2009.

In Popular Culture

Some possible future ethical problems of genetic testing were considered in the science fiction film Gattaca, the novel Next, and the science fiction anime series "Gundam Seed". Also some films which include the topic of genetic testing include, "The Island" and the "Resident Evil" series.

Limitations of Predictive Medicine

On a protein level, structure is more conserved than sequence. Therefore, in many diseases, having the faulty gene still does not necessarily mean someone will get the disease. Common, complex diseases in the wider population are affected not only by heredity, but also by external causes such as lifestyle and environment. Therefore, genes are not perfect predictors of future health; individuals with both the high risk form of the gene and those without are all candidates to get the disease. Multiple factors in the environment, particular smoking, diet and exercise, infection, and pollution; play important roles and can be more important than genetic make-up. This makes the results and risks determined by predictive medicine more difficult to quantify. Furthermore, the potential false positives or false negatives that may arise from a predictive genetic screen can cause substantial unnecessary strain on the individual.

Targeting medication to people who are genetically susceptible to a disease but do not yet show the symptoms of it can be a questionable measure. In large populations, there is concern that likely most of the people taking preventative medications would never have developed the disease anyway. Many medications carry undesirable side effects that high risk individuals must then cope with. In contrast, several populations-based prevention measures (such as encouraging healthy diets or banning tobacco advertising) carry a far lower likelihood of adverse effects and are also less expensive.

Another potential downfall of commercially available genetic testing lies within the psychological impacts of accessibility to such data. For single-gene inherited diseases, counseling and the right to refuse a test (the right "not to know") have been found to be important. However, adequate individual counseling can be difficult to employ to the potentially large proportion of the population likely to be identified as at high risk of common complex disease. Some people are vulnerable to adverse psychological reactions to genetic predictions of stigmatized or feared conditions, such as cancer or mental illness.

Ethics and Law

Predictive medicine ushers in a number of sensitive legal and ethical issues. There is a delicate balance that presides over predictive medicine and occupational health: if an employee were dismissed because he was found to be at risk of a certain chemical agent used in his workplace, would his termination be considered discrimination or an act of prevention? Several organizations believe that legislation is needed to prevent insurers and employers from using predictive genetic test results to decide who gets insurance or a job: "Ethical considerations, and legal, are fundamental to the whole issue of genetic testing.

The consequences for individuals with regard to insurance and employment are also of the greatest importance, together with the implications for stigma and discrimination." In the future, people may be required to reveal genetic predictions about their health to their employers or insurers. The grim prospect of discrimination based on a person's genetic make-up can lead to a "genetic underclass" which does not receive equal opportunity for insurance and employment. Currently in the United States, health insurers do not require applicants for coverage to undergo genetic testing. Genetic information is under the same protection of confidentiality as other sensitive health information under the Health Insurance Portability and Accountability Act (HIPAA) when health insurers come across it. In the United States, the Genetic Information Nondiscrimination Act, signed into law by President Bush on May 21, 2008; prohibits health insurers from denying coverage or charging differentials in premiums, and bars employers from making job placement or hiring/firing decisions based on individuals' genetic predispositions.

Literature Resources

In 2010, Penguin Publisher's published Outsmart Your Genes, by Brandon Colby, MD. This book is written for the lay reader and discusses the new field of predictive medicine and genetic testing from the perspective of a physician, and includes a number of pop-culture references such as to Sergei Grinkov, David Bloom, and the NFL. In the second half of Outsmart Your Genes, Dr. Colby provides specific examples of how genetic testing can be utilized to lower a person's risk of a large number of diseases, including Alzheimer's, cancer, heart disease, stroke, multiple sclerosis, asthma, and many others.

In 2010, Harper published The Language of Life: DNA and the Revolution in Personalized Medicine, by Francis S. Collins. The book is also written for the lay reader and discusses how genetic testing may be utilized to usher in personalized medicine. Dr. Collins describes general concepts of how genetics may be utilized to combat a large number of diseases, including Alzheimer's and heart disease.

Gene Theft

Gene theft or DNA theft is the act of acquiring the genetic material of another human being, often from a public place, without his or her permission. The DNA may be harvested from a wide variety of common objects such as discarded cigarettes, used coffee cups and hairbrushes. This genetic material can then be used for such purposes as establishing paternity, proving genealogical connections or even unmasking private medical conditions.

Criminal Law

Great Britain criminalized the taking of DNA without consent in 2006 at the urging of the Human Genetics Commission. Australia's legislature debated a two year jail sentence for such theft in 2008. In the United States, eight states currently have criminal or civil prohibitions on such nonconsensual appropriation of genetic materials. In Alaska, Florida, New Jersey, New York and Oregon, individuals caught swiping DNA face fines or short jail sentences. Lawsuits against "gene snatchers" are permitted in Minnesota, New Hampshire and New Mexico. In jurisdictions where such nonconsentual taking of DNA is illegal, exceptions are generally made for law enforcement.

Ethics

Many bioethicists believe that such conduct is an unethical invasion of human privacy. Professor Jacob Appel has warned that criminals may acquire the capability to copy DNA of innocent people and deposit it at crimes scenes, endangering the blameless and undermining a key tool of forensic investigation." However, others defend the appropriation of genetic material on the grounds that doing so may further human knowledge in productive ways One particularly controversial case which received widespread attention in the media was that of Derrell Teat, a wastewater coordinator, who sought to acquire without consent the DNA of a man who was allegedly the last male descendant of her great-great-great grandfather's brother. Another prominent case was a United States paternity suit involving film producer Steve Bing and billionaire investor Kirk Kerkorian.

Pseudogene

Pseudogenes are dysfunctional relatives of known genes that have lost their protein-coding ability or are otherwise no longer expressed in the cell. Although some do not have introns or promoters (these pseudogenes are copied from mRNA and incorporated into the chromosome and are called processed pseudogenes), most have some gene-like features (such as promoters, CpG islands, and splice sites), they are nonetheless considered nonfunctional, due to their lack of protein-coding ability resulting from various genetic disablements (premature stop codons, frameshifts, or a lack of transcription) or their inability to encode RNA (such as with rRNA pseudogenes). Thus the term, coined in 1977 by Jacq, et al., is composed of the prefix pseudo, which means false, and the root gene, which is the central unit of molecular genetics.

Because pseudogenes are generally thought of as the last stop for genomic material that is to be removed from the genome, they are often labeled as junk DNA. Nonetheless, pseudogenes contain fascinating biological and evolutionary histories within their sequences. This is due to a pseudogene's shared ancestry with a functional gene: in the same way that Darwin thought of two species as possibly having a shared common ancestry followed by millions of years of evolutionary divergence, a pseudogene and its associated functional gene also share a common ancestor and have diverged as separate genetic entities over millions of years.

Properties of Pseudogenes

Pseudogenes are characterized by a combination of homology to a known gene and nonfunctionality. That is, although every pseudogene has a DNA sequence that is similar to some functional gene, they are nonetheless unable to produce functional final products. Pseudogenes are quite difficult to identify and characterize in genomes, because the two requirements of homology and nonfunctionality are implied through sequence calculations and alignments rather than biologically proven.

1. Homology is implied by sequence identity between the DNA sequences of the pseudogene and parent gene. After aligning the two sequences, the percentage of identical base pairs is computed. A high sequence identity (usually between 40% and 100%) means that it is highly likely that these two sequences diverged from a common ancestral sequence (are homologous), and highly unlikely that these two sequences were independently created.
2. Nonfunctionality can manifest itself in many ways. Normally, a gene must go through several steps in going from a genetic DNA sequence to a fully functional protein: transcription, pre-mRNA processing, translation, and protein folding are all required parts of this process. If any of these steps fails, then the sequence may be considered nonfunctional. In high-throughput pseudogene identification, the most commonly identified disablements are stop codons and frameshifts, which almost universally prevent the translation of a functional protein product.

Pseudogenes for RNA genes are often easier to discover. Many RNA genes occur as multiple copy genes, and pseudogenes are identified through sequence identity and location within the region.

Types and Origin of Pseudogenes

There are three main types of pseudogenes, all with distinct mechanisms of origin and characteristic features. The classifications of pseudogenes are as follows:

1. Processed (or retrotransposed) pseudogenes. In higher eukaryotes, particularly mammals, retrotransposition is a fairly common event that has had a huge impact on the composition of the genome. For example, somewhere between 30% - 44% of the human genome consists of repetitive elements such as SINEs and LINEs. In the process of retrotransposition, a portion of the mRNA transcript of a gene is spontaneously reverse transcribed back into DNA and inserted into chromosomal DNA. Although retrotransposons usually create copies of themselves, it has been shown in an in vitro system that they can create retrotransposed copies of random genes, too. Once these pseudogenes are inserted back into the genome, they usually contain a poly-A tail, and usually have had their introns spliced out; these are both hallmark features of cDNAs. However, because they are derived from a mature mRNA product, processed pseudogenes also lack the upstream promoters of normal genes; thus, they are considered "dead on arrival", becoming non-functional pseudogenes immediately upon the retrotransposition event. However, occasionally these insertions contribute exons to existing genes and usually via alternatively spliced transcripts. A further characteristic of processed pseudogenes is common truncation of the 5' end relative to the parent sequence, which is a result of the relatively non-processive retrotransposition mechanism that creates processed pseudogenes.
2. Non-processed (or duplicated) pseudogenes. Gene duplication is another common and important process in the evolution of genomes. A copy of a functional gene may arise as a result of a gene duplication event and subsequently acquire mutations that cause it to become nonfunctional. Duplicated pseudogenes usually have all the same characteristics of genes, including an intact exon-intron structure and promoter sequences. The

loss of a duplicated gene's functionality usually has little effect on an organism's fitness, since an intact functional copy still exists. According to some evolutionary models, shared duplicated pseudogenes indicate the evolutionary relatedness of humans and the other primates.

3. Disabled genes, or unitary pseudogenes. Various mutations can stop a gene from being successfully transcribed or translated, and a gene may become nonfunctional or deactivated if such a mutation becomes fixed in the population. This is the same mechanism by which non-processed genes become deactivated, but the difference in this case is that the gene was not duplicated before becoming disabled. Normally, such gene deactivation would be unlikely to become fixed in a population, but various population effects, such as genetic drift, a population bottleneck, or in some cases, natural selection, can lead to fixation. The classic example of a unitary pseudogene is the gene that presumably coded the enzyme L-gulono-ã-lactone oxidase (GULO) in primates. In all mammals studied besides primates (except guinea pigs), GULO aids in the biosynthesis of Ascorbic acid (vitamin C), but it exists as a disabled gene (GULOP) in humans and other primates. Another interesting and more recent example of a disabled gene, which links the deactivation of the caspase 12 gene (through a nonsense mutation) to positive selection in humans.

Pseudogenes can complicate molecular genetic studies. For example, a researcher who wants to amplify a gene by PCR may simultaneously amplify a pseudogene that shares similar sequences. This is known as PCR bias or amplification bias. Similarly, pseudogenes are sometimes annotated as genes in genome sequences.

Processed pseudogenes often pose a problem for gene prediction programs, often being misidentified as real genes or exons. It has been proposed that identification of processed pseudogenes can help improve the accuracy of gene prediction methods.

It has also been shown that the parent sequences that give rise to processed pseudogenes lose their coding potential faster than those giving rise to non-processed pseudogenes.

Functional Pseudogenes?

By definition, pseudogenes lack a function. However, the classification of pseudogenes generally relies on computational analysis of genomic sequences using complex algorithms. This has led to the incorrect identification of pseudogenes. For example the functional, chimeric gene jingwei in Drosophila was once thought to be a processed pseudogene. It has been established that quite a few pseudogenes can go through the process of transcription, either if their own promoter is still intact or in some cases using the promoter of a nearby gene; this expression of pseudogenes also appears to be tissue-specific.

In 2003, Hirotsune et al. identified a retrotransposed pseudogene whose transcript purportedly plays a trans-regulatory role in the expression of its homologous gene, Makorin1 (MKRN1) (RING finger domain and ubiquitin ligases), and suggested this as a general model under which pseudogenes may play an important biological role.

Other researchers have since hypothesized similar roles for other pseudogenes. A bioinformatics analysis has shown that processed pseudogenes can be inserted into introns of annotated genes and be incorporated into alternatively spliced transcripts. Hirotsune's report prompted two molecular biologists to carefully review scientific literature on the subject of pseudogenes. To the surprise of many, they found a number of examples in which pseudogenes play a role in gene regulation and expression, forcing Hirotsune's group to rescind their claim that they were the first to identify pseudogene function. Furthermore, the original findings of Hirotsune et al. concerning Makorin1 have recently been strongly contested; thus, the possibility that some pseudogenes could have important biological functions was disputed. Additionally, University of Chicago and University of Cincinnati scientists reported in 2002 that a processed pseudogene called phosphoglycerate mutase 3 (PGAM3P) actually produces a functional protein.

Two 2008 publications in Nature discuss that some endogenous siRNAs are derived from pseudogenes, and thus some pseudogenes play a role in regulating protein-coding transcripts. In June 2010, Nature published an article showing the mRNA levels of tumour suppressor PTEN and oncogenic KRAS is affected by their pseudogenes PTENP1 and KRAS1P. This discovery demonstrated an miRNA decoy function for pseudogenes and identified their transcripts as biologically active units in tumor biology; thus attributing a novel biological role to expressed pseudogenes, as they can regulate coding gene expression, and reveal a non-coding function for mRNAs in disease progression.

Microarray Technology: Past, Present and Future

Molecular Biology research evolves through the development of the technologies used for carrying them out. It is not possible to research on a large number of genes using traditional methods. DNA

Microarray is one such technology which enables the researchers to investigate and address issues which were once thought to be non traceable. One can analyze the expression of many genes in a single reaction quickly and in an efficient manner. DNA Microarray technology has empowered the scientific community to understand the fundamental aspects underlining the growth and development of life as well as to explore the genetic causes of anomalies occurring in the functioning of the human body.

A typical microarray experiment involves the hybridization of an mRNA molecule to the the DNA template from which it is originated. Many DNA samples are used to construct an array. The amount of mRNA bound to each site on the array indicates the expression level of the various genes. This number may run in thousands. All the data is collected and a profile is generated for gene expression in the cell.

Microarray Technique

An array is an orderly arrangement of samples where matching of known and unknown DNA samples is done based on base pairing rules. An array experiment makes use of common assay systems such as microplates or standard blotting membranes. The sample spot sizes are typically less than 200 microns in diameter usually contain thousands of spots.

Thousands of spotted samples known as probes (with known identity) are immobilized on a solid support (a microscope glass slides or silicon chips or nylon membrane). The spots can be DNA, cDNA, or oligonucleotides. These are used to determine complementary binding of the unknown sequences thus allowing parallel analysis for gene expression and gene discovery.

An experiment with a single DNA chip can provide information on thousands of genes simultaneously. An orderly arrangement of the probes on the support is important as the location of each spot on the array is used for the identification of a gene.

Types of Microarrays

Depending upon the kind of immobilized sample used construct arrays and the information fetched, the Microarray experiments can be categorized in three ways:

1. Microarray expression analysis: In this experimental setup, the cDNA derived from the mRNA of known genes is immobilized. The sample has genes from both the normal as well as the diseased tissues. Spots with more more intensity are obtained

for diseased tissue gene if the gene is over expressed in the diseased condition. This expression pattern is then compared to the expression pattern of a gene responsible for a disease.

2. Microarray for mutation analysis: For this analysis, the researchers use gDNA. The genes might differ from each other by as less as a single nucleotide base.

A single base difference between two sequences is known as Single Nucleotide Polymorphism (SNP) and detecting them is known as SNP detection.

3. Comparative Genomic Hybridization: It is used for the identification in the increase or decrease of the important chromosomal fragments harboring genes involved in a disease.

Applications of Microarrays

Gene discovery: DNA Microarray technology helps in the identification of new genes, know about their functioning and expression levels under different conditions.

Disease diagnosis: DNA Microarray technology helps researchers learn more about different diseases such as heart diseases, mental illness, infectious disease and especially the study of cancer. Until recently, different types of cancer have been classified on the basis of the organs in which the tumors develop. Now, with the evolution of microarray technology, it will be possible for the researchers to further classify the types of cancer on the basis of the patterns of gene activity in the tumor cells. This will tremendously help the pharmaceutical community to develop more effective drugs as the treatment strategies will be targeted directly to the specific type of cancer.

Drug discovery: Microarray technology has extensive application in Pharmacogenomics. Pharmacogenomics is the study of correlations between therapeutic responses to drugs and the genetic profiles of the patients. Comparative analysis of the genes from a diseased and a normal cell will help the identification of the biochemical constitution of the proteins synthesized by the diseased genes. The researchers can use this information to synthesize drugs which combat with these proteins and reduce their effect.

Toxicological research: Microarray technology provides a robust platform for the research of the impact of toxins on the cells and their passing on to the progeny. Toxicogenomics establishes correlation between responses to toxicants and the changes in the genetic profiles of the cells exposed to such toxicants.

GEO

In the recent past, microarray technology has been extensively used by the scientific community. Consequently, over the years, there has been a lot of generation of data related to gene expression. This data is scattered and is not easily available for public use. For easing the accessibility to this data, the National Center for Biotechnology Information (NCBI) has formulated the Gene Expression Omnibus or GEO.

It is a data repository facility which includes data on gene expression from varied sources.

Microarray Probe Design Parameters

Table:

For 25-35 mers

Parameter	Minimum Value	Maximum Value	Default Value	Unit
Probe Length	*10*	*99*	*30*	*bases*
Probe Length tolerance	*0*	*15*	*3*	
Probe Target Tm	*40*	*99*	*63*	*°C*
Probe Tm Tolerance (+)	*0.1*	*99*	*5*	
Hairpin Max Δ *G*	*0.1*	*99.9*	*4*	*Kcal/mol*
Self Dimer Δ *G*	*0.1*	*99.9*	*7*	*Kcal/mol*
Run/Repeat	*2*	*99*	*4*	*bases*

Table:

Parameter	Minimum Value	Maximum Value	Default Value	Unit
Probe Length	10	99	40	bases
Probe Length tolerance	0	15	3	
Probe Target Tm	40	99	70	°C
Probe Tm Tolerance (+)	0.1	99	5	
Hairpin Max Δ G	0.1	99.9	6	Kcal/mol
Self Dimer Δ G	0.1	99.9	8	Kcal/mol
Run/Repeat	2	99	5	bases

Table:

Parameter	Minimum Value	Maximum Value	Default Value	Unit
Probe Length	10	99	70	bases
Probe Length tolerance	0	15	3	
Probe Target Tm	40	99	75	°C
Probe Tm Tolerance				

(+/- above)	0.1	99	5	
Hairpin Max Δ G	0.1	99.9	6	Kcal/mol
Self Dimer Δ G	0.1	99.9	8	Kcal/mol
Run/Repeat	2	99	6	bases

Molecular diagnostic tests typically analyze key protein, DNA, or RNA markers to characterize diseases. Immunohistochemistry, Western blotting, DNA sequencing, polymerase chain reaction (PCR), Southern blotting, Northern blotting, and quantitative reverse-transcription PCR (qRT-PCR) are molecular techniques that have been effectively used in clinical diagnostic laboratories, where broadly trained technologists perform the tests rather than researchers with specific methodological expertise. However, more-parallel molecular tests that involve simultaneous analysis of tens to hundreds of markers within a clinical sample will soon be required to allow maximum translation of postgenomic research to patient care.

Many basic research projects have focused on methods to measure the behavior of the approximately 30,000 genes in the human genome. This research has been done at different levels within the gene, measuring both genotypic and phenotypic changes. For example, parallel analysis has been applied at the DNA level to analyze sequence polymorphisms that distinguish closely related human alleles or microbial genes; at the level of DNA methylation patterns; at the mRNA level to analyze gene expression patterns; and at the protein level.

With the development of microarray-based tests for diagnosing a wide range of diseases, the ability to gain more information from a limited clinical sample by using highly parallel expression-analysis techniques has emerged. This technology's greatest potential exists for those diseases in which histological differentiation between disease entities is difficult and the consequences of inappropriate treatment can be serious. In the research environment, analyzing sets of gene products with microarrays, rather than individual proteins or RNAs, has been shown to provide effective diagnoses. However, improving high-throughput diagnostic methods will require the continued development of new second-generation custom microarrays. This article describes the challenges in developing this technology and the custom microarray techniques that are currently available.

Microarrays as Transcription Analysis Tools

High-density DNA (HD DNA) microarrays have been useful in the semi-quantitative, or comparative, analysis of thousands of individual messenger RNA (mRNA) species. Experiments using HD

DNA microarrays suggest that new tests measuring the expression of tens to hundreds of genes will be better able to characterize disease states than is currently possible using histopathology and other existing molecular methods. As stated above, clinical application of this DNA microarray-based research will require further development of sophisticated new second-generation custom DNA microarray platforms that are more suitable for routine clinical laboratory testing.

HD DNA microarrays are small devices that hold tens of thousands of gene capture probes at known positions on a solid surface. One of the most widely used platforms is the arrays by Affymetrix Inc. (Santa Clara, CA), in which DNA oligonucleotides are chemically synthesized in situ on the array surface using masked photolithography. Another widely used platform is the cDNA array, in which PCR products are generated separately and placed on the array surface using a robotic arrayer. Thousands of laboratories have used these two platforms to catalog the expression of large numbers of genes in normal and diseased cells and tissues.

When transitioning from a normal to a diseased state, cells and tissues undergo a variety of histological and biochemical changes. These changes reflect in large part the dysregulation in the expression of genes. The quantities and types of proteins that are expressed from their respective genes determine the normal and abnormal functions of cells. When genes are expressed, their DNA sequences are first copied into temporary molecules, or mRNAs. These mRNAs are then translated into proteins that perform the functions and generate the structures that allow cells to conduct their biological activities.

Researchers can take advantage of this flow of genetic information from DNA to mRNAs to proteins by purifying populations of all mRNAs expressed in specimens, synthesizing fluorescently labeled cDNA copies of the mRNAs in vitro, and hybridizing the labeled cDNAs to a microarray. The intensity of the fluorescent signal at each gene-specific probe on the microarray reflects the level of mRNA expressed from that gene in the specimen. By measuring the types and quantities of tens of thousands of individual mRNAs expressed in samples, researchers can infer information regarding the state of gene expression, the approximate quantities and types of proteins present, and the biological behavior of the sample.

For several decades, alternative methods have been available for identifying genes whose expression patterns are altered in disease. However, HD DNA microarrays have increased research scientists' productivity in their gene- hunting efforts. As little as five years ago, identifying even a few differentially expressed genes may have taken several years and cost tens of thousands of dollars. Today, HD DNA microarrays can identify 10 times that number of genes in a few months and at a tenth of the cost.

Advances in Lymphoma Diagnostics

Recent studies on the biology of non-Hodgkin's lymphomas (NHL) illustrate some of the diagnostic advances made possible by HD DNA microarrays. In 2000, it is estimated that 54,900 patients were newly diagnosed with NHLs and 22,553 died of these diseases. Although there are many NHL subtypes, the diffuse large B-cell lymphoma (DLBCL) subtype is the most common, comprising 30% of all cases.

Despite the complexity of this classification system, it is increasingly recognized that the NHL categories themselves encompass multiple disease subtypes with distinct molecular defects and clinical outcomes. For example, while 35–40% of patients diagnosed with DLBCL can be cured with chemotherapy, the remaining 60–65% die of this disease. Numerous attempts to distinguish patients who are likely to respond to chemotherapy from those unlikely to respond have been unsuccessful. In addition, these DLBCL tumors are indistinguishable using classical histopathology and currently available molecular markers.

Researchers have approached this problem by using HD cDNA microarrays to profile gene expressions in DLBCL. These researchers identified genes that classified DLBCL specimens into distinct subclasses. One subclass, germinal-center B-cell-like DLBCL, showed an expression pattern similar to mature germinal-center B-cells and portended a more favorable prognosis with a 60% five-year survival rate. On the contrary, two other DLBCL subclasses, activated B-cell-like DLBCL and type-3 DLBCL, showed much less favorable prognoses with a 35% five-year survival rate.

Using Microarrays for Personalized Medicine

Research studies have predicted that in the near future, it may be possible to determine which patients might benefit from more-aggressive therapies, such as bone marrow transplants or alternative chemotherapy regimens, when DLBCL is diagnosed. HD DNA microarray studies have also highlighted molecular pathways important in subclasses of cancer.

For example, researchers have observed that several genes in the activated B-cell subtype of DLBCL were downstream targets of the NFkB transcription factor. These investigators demonstrated that

NFkB activity is indeed higher in this DLBCL subtype, suggesting that drugs specifically targeting the NFkB pathway may be effective in treating these tumors.

Several recent papers have suggested that gene expression profiling of tumor specimens may be useful for preselecting patients who may benefit from drug treatment. It may also be useful to examine gene expression profiles of cancers following chemotherapy in order to determine whether the tumors are responding to treatment. The goal of this approach is personalized medicine, in which detailed patient-specific molecular information would be assayed to predict an effective therapy.

Microarrays as Genotyping Tools for Infectious Diseases

Despite the abundance of protein diagnostics, diagnosing diseases by examining the qualitative presence of specific nucleic acid sequences has been adopted for infectious diseases. While the amplification of gene sequences through PCR or other similar methods can indicate the presence or absence of a pathogen genomic sequence, nucleic acid hybridization is almost always necessary to confirm that the amplification is sequence-specific. Suitably designed microarrays are versatile tools for this function since the hybridization data from an array can identify the pathogen genomic sequence in a more rapid and cost-effective manner than DNA sequencing.

The sudden emergence of new strains of pathogens has brought the need for rapid nucleic acid diagnostic tools to the forefront. Although customizable nucleic acid testing systems for the rapid development of diagnostics are still new and emerging, the urgent medical need for high-speed diagnostics is steadily increasing. For example, the recent outbreak of severe acute respiratory syndrome (SARS) and its rapid spread left no time for generating an antibody-based diagnostic test. Initial identification of the virus was based on symptom presentation and high-resolution electron microscopy of sputum samples. However, gene sequencing of the emergent SARS strains permitted rapid identification of the virus by using a nucleic acid amplification and hybridization/sequencing approach.

Monitoring how the genotype of viruses evolves is critical for understanding the pathology of a disease as it spreads throughout a population. Viruses such as influenza, poliovirus, rhinoviruses, and HIV are known to have high rates of mutation, in some cases even causing attenuated strains that are developed for vaccines to revert to viruses with a wild-type phenotype. Although it remains possible to generate antibodies for all of these pathogen strains, identifying strains rapidly is more easily achieved using microarray tools developed specifically for this purpose.

Second-Generation Custom DNA Microarrays

Many researchers agree that in some form, DNA microarrays will be used as diagnostic tools in the near future. The first prospective trials are under way to determine whether microarrays can predict the proper clinical course for patients.

At the same time, FDA is anticipating working on the regulatory aspects of microarray-based diagnostic tests. However, extensive work will be necessary to develop microarray platforms that give reproducible results in different laboratories, which will be required for clinical applications. In addition, the basic parameters of microarray technology will also need to be established, including content selection (i.e., which oligonucleotides will be used), data analysis, data interpretation, and uniform sample preparation and assay conditions.

The use of a variety of generic whole-genome HD DNA microarrays in the research setting has led to important new basic research findings. However, these platforms are not suitable for use in clinical diagnostic laboratories for a number of reasons.

First, ownership claims for probes to particular gene sequences can burden the expensive diagnostic development process before it has even begun. Second, the lack of sequence flexibility of the nucleic acid capture probes on the microarrays makes it difficult to be sure that the best capture probe has been identified. In other words, there is no chance to improve rapidly and iteratively the performance of the capture probe to its target.

Third, many genes have significant polymorphisms or splice variants, which may or may not have been correctly represented on the array. The content arrays are based on the human genomic information from a comparatively small set of human genomic sequences. Finally, nearly all of these products are built around gene expression type applications, which are not necessarily the only approach for developing diagnostic tools.

In contrast, customizable microarrays give diagnostics developers complete freedom to identify rapid nucleic acid probes for specific applications, such as key changes in gene-expression approach, disease-related polymorphisms, patterns in DNA methylation analysis, or resequencing applications for infectious diseases. Although a number of technologies exist for making custom arrays, there are significant differences in the techniques used to fabricate the arrays, which have an impact on the length of time, cost, and complexity in developing nucleic acid diagnostics.

For example, the Affymetrix system uses specially made chromium photomasks from the semiconductor industry to pattern the synthesis of nucleic acids. Although this method has been widely adopted in research scenarios, an important limiting factor concerns poor stepwise yields of light-based deprotection. Typically, nucleotides are used for light-based deprotection systems. However, while short oligonucleotides may work for some applications, they may not be ideal for developing many nucleic acid diagnostic tools.

Similarly, spotted cDNA arrays encounter problems with spot nonuniformity and misidentification of spots. These problems limit the quality of data derived from high-density microarrays, making it difficult to distinguish experimental noise from biologically meaningful measurements of gene expression. For example, a significant percentage of genes that are identified as being differentially expressed from specimen types using high-density microarrays cannot be confirmed using the more-laborious gold standard technology of qRT-PCR. This is particularly true for genes that are expressed at low levels.

In addition, several recent papers have described similar experiments that were performed on both the Affymetrix and spotted cDNA platforms and produced different lists of differentially expressed genes. If the gene expression patterns were genuine, researchers should be able to identify the same genes regardless of which platform is used. This noncorrelation of data across platforms suggests that care should be taken to understand the technical limitations of these platforms.

An alternative technology that addresses the key issues of turnaround time, interplatform reproducibility, and oligonucleotide quality is the CustomArray, a semiconductor-based technology by CombiMatrix Corp. (Mukilteo, WA).

This technology uses the electrochemical generation of acid at microelectrodes, rather than light-based photochemistry, to pattern nucleic acid synthesis. With a 16–36-hour turnaround time, this system can synthesize oligonucleotides of high quality and long lengths on an array format.

This technology enables testing a wide range of diagnostic approaches, including screening genotypic information from both virus and host cells concurrently, using multiple assay types on one chip, and employing variable oligonucleotide lengths and non standard chemistries. This technology also benefits from employing established manufacturing processes that have been developed in the semiconductor industry. This is an important consideration when determining how a diagnostic device can be manufactured in large volumes at a reasonable cost.

Conclusion

An ideal custom DNA microarray platform should allow a user complete freedom in defining which genes are assayed using the array. The microarray platform should also be compatible with computer-based probe design software and molecular methods so that measurements can be performed with high sensitivity and specificity, thereby making independent validation of the assay unnecessary. The use of manufacturing techniques that allow miniaturization and massive scale-up and result in low costs and minimal biological sample requirements should contribute to a well-engineered microarray platform. Because of these factors, there should be standardization among various clinical laboratories using microarrays, such that results from different laboratories can be directly compared. This is particularly important under existing and proposed regulations that give the Centers for Disease Control and Prevention (Atlanta) and FDA shared responsibility for monitoring molecular diagnostic testing.

Dominant and Recessive Genetic Diseases in Humans

In humans, many genetic traits or diseases are classified simply as "dominant" or "recessive." Especially with respect to so-called recessive diseases, this can oversimplify the underlying molecular basis and lead to misunderstanding of the nature of dominance. For example, the genetic disease phenylketonuria (PKU) results from any of a large number (>60) of alleles at the gene locus for the enzyme phenylalanine hydroxylase (PAH). Many of these alleles produce little or no PAH, as a result of which the substrate phenylalanine and its metabolic byproducts accumulate in the central nervous system and can cause severe mental retardation if untreated. The genotypes and phenotypic consequences of interactions among three alleles are shown in the following table:

Table:

Genotype	PAH activity	[phe] conc	PKU ?
AA	*100%*	*60 uM*	*No*
AB	*30%*	*120 uM*	*No*
CC	*5%*	*200 ~ 300 uM*	*Hyperphenylalanemia*
BB	*0.3%*	*600 ~ 2400 uM*	*Yes*

In unaffected persons homozygous for a standard functional allele (AA), PAH activity is standard (100%), and the concentration of phenylalanine in the blood [phe] is about 60 uM. In untreated persons homozygous for one of the PKU alleles (BB), PAH activity is close to zero, [phe] ten to forty times standard, and the individual manifests PKU.

In the AB heterozygote, PAH activity is only 30% (not 50%) of standard, blood [phe] is elevated two-fold, and the person does not manifest PKU. Thus, the A allele is dominant to the B allele with respect to PKU, but the B allele is incompletely dominant to the A allele with respect to its molecular effect, determination of PAH activity level (0.3% < 30% << 100%). Finally, the A allele is an incomplete dominant to B with respect to [phe], as 60 uM < 120 uM << 600 uM. Note once more that it is irrelevant to the question of dominance that the recessive allele produces a more extreme [phe] phenotype. For a third allele C, a CC homozygote produces a very small amount of PAH enzyme, which results in a somewhat elevated level of [phe] in the blood, a condition called hyperphenylalanemia, which does not result in mental retardation.

That is, the dominance relationships of any two alleles may vary according to which aspect of the phenotype is under consideration. It is typically more useful to talk about the phenotypic consequences of the allelic interactions involved in any genotype, rather than to try to force them into dominant and recessive categories.

History

The concept of dominance was first described by the "Father of Genetics," Gregor Mendel, in the 1860s. Mendel observed that, for a variety of traits of garden peas having to do with the appearance of seeds, seed pods, and plant appearance, there occurred two discrete phenotypes: round vs wrinkled, or yellow vs green seeds, red vs white flowers, tall vs short plants, and so on. When bred separately, the plants always produced the same phenotypes, generation after generation. However, when lines with different phenotypes were crossed (interbred), one and only one of the parental phenotypes showed up in the offspring: green, or round, or red, or tall, and so on. However, when these hybrid plants were crossed, the offspring plants showed the two original phenotypes, in a characteristic 3:1 ratio, with the more common type having the phenotype of the parental hybrid plants. Mendel reasoned that each of the parents in the first cross were homozygotes for different alleles (AA and aa), that each contributed one allele to the offspring, such that all of these hybrids were heterozygotes (Aa), and that one of the two alleles in the hybrid cross dominated expression of the other: A masked a.

The final cross between two heterozygotes (Aa X Aa) would produce AA, Aa, and aa offspring in a 1:2:1 genotype ratio with the first three classes showing the "A" phenotype, and the last showing the "a" phenotype, thereby producing the 3:1 phenotype ratio.

Mendel did not use the terms gene, allele, phenotype, genotype, homozygote, and heterozygote, all of which were introduced afterward. He did introduce the notation of capital and lowercase letters for dominant and recessive alleles, respectively, still in use today.

Mitochondrial DNA

Mitochondrial DNA (mtDNA) is the DNA located in organelles called mitochondria, structures within eukaryotic cells that convert the chemical energy from food into a form that cells can use, adenosine triphosphate (ATP).

Most other DNA present in eukaryotic organisms is found in the cell nucleus. Mitochondrial DNA can be regarded as the smallest chromosome, and was the first significant part of the human genome to be sequenced. In most species, including humans, mtDNA is inherited solely from the mother.

The DNA sequence of mtDNA has been determined from a large number of organisms and individuals (including some organisms that are extinct), and the comparison of those DNA sequences represents a mainstay of phylogenetics, in that it allows biologists to elucidate the evolutionary relationships among species. It also permits an examination of the relatedness of populations, and so has become important in anthropology and field biology.

Replication

mtDNA is replicated by the DNA polymerase gamma complex which is composed of a 140 kDa catalytic DNA polymerase encoded by the POLG gene and a 55 kDa accessory subunit encoded by the POLG2 gene.

During embryogenesis, replication of mtDNA is strictly down-regulated from the fertilized oocyte through the preimplantation embryo. At the blastocyst stage, the onset of mtDNA replication is specific to the cells of the trophectoderm. In contrast, the cells of the inner cell mass restrict mtDNA replication until they receive the signals to differentiate to specific cell types.

Origin

Nuclear and mitochondrial DNA are thought to be of separate evolutionary origin, with the mtDNA being derived from the circular genomes of the bacteria that were engulfed by the early ancestors of today's eukaryotic cells. This theory is called the endosymbiotic theory. Each mitochondrion is estimated to contain 2-10 mtDNA copies. In the cells of extant organisms, the vast majority of the proteins present in the mitochondria (numbering approximately 1500 different types in mammals) are coded for by nuclear DNA, but the genes for some of them, if not most, are thought to have originally been of bacterial origin, having since been transferred to the eukaryotic nucleus during evolution.

Mitochondrial Inheritance

In most multicellular organisms, mtDNA is inherited from the mother (maternally inherited). Mechanisms for this include simple dilution (an egg contains 100,000 to 1,000,000 mtDNA molecules, whereas a sperm contains only 100 to 1000), degradation of sperm mtDNA in the fertilized egg, and, at least in a few organisms, failure of sperm mtDNA to enter the egg. Whatever the mechanism, this single parent (uniparental) pattern of mtDNA inheritance is found in most animals, most plants and in fungi as well.

Female Inheritance

In sexual reproduction, mitochondria are normally inherited exclusively from the mother. The mitochondria in mammalian sperm are usually destroyed by the egg cell after fertilization. Also, most mitochondria are present at the base of the sperm's tail, which is used for propelling the sperm cells. Sometimes the tail is lost during fertilization. In 1999 it was reported that paternal sperm mitochondria (containing mtDNA) are marked with ubiquitin to select them for later destruction inside the embryo. Some in vitro fertilization techniques, particularly injecting a sperm into an oocyte, may interfere with this.

The fact that mitochondrial DNA is maternally inherited enables researchers to trace maternal lineage far back in time. (Y-chromosomal DNA, paternally inherited, is used in an analogous way to trace the agnate lineage.)

This is accomplished in humans by sequencing one or more of the hypervariable control regions (HVR1 or HVR2) of the mitochondrial DNA, as with a genealogical DNA test. HVR1 consists of about 440 base pairs. These 440 base pairs are then compared to the control regions of other individuals (either specific people or subjects in a database) to determine maternal lineage. Most often, the comparison is made to the revised Cambridge Reference Sequence. Vilà et al. have published studies tracing the matrilineal descent of domestic dogs to wolves. The concept of the Mitochondrial Eve is based on the same type of analysis, attempting to discover the origin of humanity by tracking the lineage back in time.

Because mtDNA is not highly conserved and has a rapid mutation rate, it is useful for studying the evolutionary relationships - phylogeny - of organisms.

Biologists can determine and then compare mtDNA sequences among different species and use the comparisons to build an evolutionary tree for the species examined.

Because mtDNA is transmitted from mother to child (both male and female), it can be a useful tool in genealogical research into a person's maternal line.

Male Inheritance

It has been reported that mitochondria can occasionally be inherited from the father in some species such as mussels. Paternally inherited mitochondria have additionally been reported in some insects such as fruit flies, honeybees, and periodical cicadas.

Evidence supports rare instances of male mitochondrial inheritance in some mammals as well. Specifically, documented occurrences exist for mice, where the male-inherited mitochondria was subsequently rejected. It has also been found in sheep, and in cloned cattle.

It has been found in a single case in a human male. While many of these cases involve cloned embryos or subsequent rejection of the paternal mitochondria, others document in vivo inheritance and persistence under lab conditions.

Structure

In humans (and probably in metazoans in general), 100-10,000 separate copies of mtDNA are usually present per cell (egg and sperm cells are exceptions). In mammals, each double-stranded circular mtDNA molecule consists of 15,000-17,000 base pairs.

The two strands of mtDNA are differentiated by their nucleotide content with the guanine rich strand referred to as the heavy strand, and the cytosine rich strand referred to as the light strand. The heavy strand encodes 28 genes, and the light strand encodes 9 genes for a total of 37 genes. Of the 37 genes, 13 are for proteins (polypeptides), 22 are for transfer RNA (tRNA) and two are for the small and large subunits of ribosomal RNA (rRNA). This pattern is also seen among most metazoans, although in some cases one or more of the 37 genes is absent and the mtDNA size range is greater.

Even greater variation in mtDNA gene content and size exists among fungi and plants, although there appears to be a core subset of genes that are present in all eukaryotes (except for the few that have no mitochondria at all).

Some plant species have enormous mtDNAs (as many as 2,500,000 base pairs per mtDNA molecule) but, surprisingly, even those huge mtDNAs contain the same number and kinds of genes as related plants with much smaller mtDNAs.

The genome of the mitochondrion of the cucumber (Cucumis sativus) is three entirely or largely autonomous circular chromosomes (lengths 1556, 84 and 45 kilobases).

Genes

Note: The gene content of mitochondrial DNA varies among species. Information in this section is specific to human mitochondrial DNA

Electron Transport Chain

The mitochondrial genome contains 13 protein-coding genes. Many of these genes encode proteins of the electron transport chain:

Table:

Category	Genes
NADH dehydrogenase (complex I)	*MT-ND1, MT-ND2, MT-ND3, MT-ND4, MT-ND4L, MT-ND5, MT-ND6*
Coenzyme Q - cytochrome c reductase/Cytochrome b (complex III)	*MT-CYB*
cytochrome c oxidase (complex IV)	*MT-CO1, MT-CO2, MT-CO3*
ATP synthase	*MT-ATP6, MT-ATP8*

rRNA

Mitochondrial rRNA is encoded by MT-RNR1 (12S) and MT-RNR2 (16S).

tRNA

Table: The following genes encode tRNA

Amino Acid	3-Letter	1-Letter	MT DNA
Alanine	*Ala*	*A*	*MT-TA*
Arginine	*Arg*	*R*	*MT-TR*
Asparagine	*Asn*	*N*	*MT-TN*
Aspartic acid	*Asp*	*D*	*MT-TD*
Cysteine	*Cys*	*C*	*MT-TC*
Glutamic acid	*Glu*	*E*	*MT-TE*
Glutamine	*Gln*	*Q*	*MT-TQ*
Glycine	*Gly*	*G*	*MT-TG*
Histidine	*His*	*H*	*MT-TH*
Isoleucine	*Ile*	*I*	*MT-TI*
Leucine	*Leu*	*L*	*MT-TL1, MT-TL2*
Lysine	*Lys*	*K*	*MT-TK*
Methionine	*Met*	*M*	*MT-TM*
Phenylalanine	*Phe*	*F*	*MT-TF*
Proline	*Pro*	*P*	*MT-TP*
Serine	*Ser*	*S*	*MT-TS1, MT-TS2*
Threonine	*Thr*	*T*	*MT-TT*
Tryptophan	*Trp*	*W*	*MT-TW*
Tyrosine	*Tyr*	*Y*	*MT-TY*
Valine	*Val*	*V*	*MT-TV*

Mutations

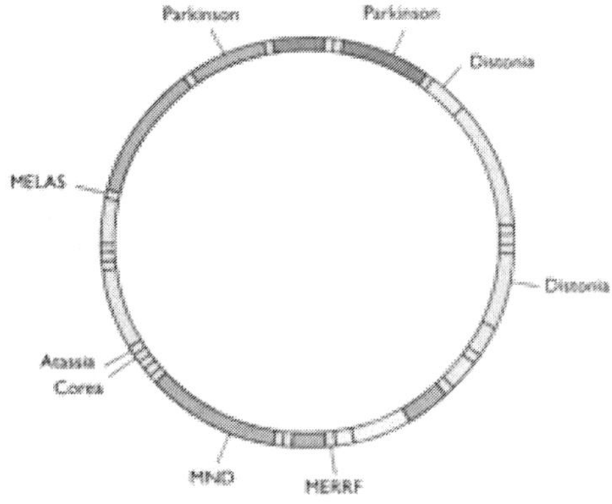

Figure: The involvement of mitochondrial DNA in several human diseases.

Susceptibility

mtDNA is particularly susceptible to reactive oxygen species generated by the respiratory chain due to its proximity. Though mtDNA is packaged by proteins and harbors significant DNA repair capacity, these protective functions are less robust than those operating on nuclear DNA and therefore thought to contribute to enhanced susceptibility of mtDNA to oxidative damage. The outcome of mutation in mtDNA may be alteration in the coding instructions for some proteins, which may have an effect on organism metabolism and/or fitness.

Genetic Illness

Mutations of mitochondrial DNA can lead to a number of illnesses including exercise intolerance and Kearns-Sayre syn drome (KSS), which causes a person to lose full function of heart, eye, and muscle movements. Some evidence suggests that they might be major contributors to the aging process and age-associated pathologies.

Use in Disease Diagnosis

Recently a mutation in mtDNA has been used to help diagnose prostate cancer in patients with negative prostate biopsy.

Use in Identification

In humans, mitochondrial DNA spans 16,569 DNA building blocks (base pairs), representing a fraction of the total DNA in cells. Unlike nuclear DNA, which is inherited from both parents and in which genes are rearranged in the process of recombination, there is usually no change in mtDNA from parent to offspring. Although mtDNA also recombines, it does so with copies of itself within the same mitochondrion. Because of this and because the mutation rate of animal mtDNA is higher than that of nuclear DNA, mtDNA is a powerful tool for tracking ancestry through females (matrilineage) and has been used in this role to track the ancestry of many species back hundreds of generations.

Human mtDNA can also be used to help identify individuals. Forensic laboratories occasionally use mtDNA comparison to identify human remains, and especially to identify older unidentified skeletal remains. Although unlike nuclear DNA, mtDNA is not specific to one individual, it can be used in combination with other evidence (anthropological evidence, circumstantial evidence, and the like) to establish identification. mtDNA is also used to exclude possible matches between missing persons and unidentified remains. Many researchers believe that mtDNA is better suited to identification of older skeletal remains than nuclear DNA because the greater number of copies of mtDNA per cell increases the chance of obtaining a useful sample, and because a match with a living relative is possible even if numerous maternal generations separate the two.

American outlaw Jesse James's remains were identified using a comparison between mtDNA extracted from his remains and the mtDNA of the son of the female-line great-granddaughter of his sister. Similarly, the remains of Alexandra Feodorovna (Alix of Hesse), last Empress of Russia, and her children were identified by comparison of their mitochondrial DNA with that of Prince Philip, Duke of Edinburgh, whose maternal grandmother was Alexandra's sister Victoria of Hesse.

Similarly to identify Emperor Nicholas II remains his mitochondrial DNA was compared with that of James Carnegie, 3rd Duke of Fife, whose maternal great-grandmother Alexandra of Denmark (Queen Alexandra) was sister of Nicholas II mother Dagmar of Denmark (Empress Maria Feodorovna). The low effective population size and rapid mutation rate (in animals) makes mtDNA useful for assessing genetic relationships of individuals or groups within a species and also for identifying and quantifying the phylogeny (evolutionary relationships; phylogenetics) among different species, provided they are not too distantly related. To do this, biologists determine and then compare the mtDNA sequences from different individuals or species.

Data from the comparisons is used to construct a network of relationships among the sequences, which provides an estimate of the relationships among the individuals or species from which the mtDNAs were taken. This approach has limits that are imposed by the rate of mtDNA sequence change. In animals, the high mutation rate makes mtDNA most useful for comparisons of individuals within species and for comparisons of species that are closely or moderately-closely related, among which the number of sequence differences can be easily counted. As the species become more distantly related, the number of sequence differences becomes very large; changes begin to accumulate on changes until an accurate count becomes impossible.

History

Mitochondrial DNA was discovered in the 1960s by Margit M. K. Nass and Sylvan Nass by electron microscopy as DNase-sensitive thread inside mitochondria, and by Ellen Haslbrunner, Hans Tuppy and Gottfried Schatz by biochemical assays on highly purified mitochondrial fractions.

Sex Linkage

Sex linkage is the phenotypic expression of an allele related to the chromosomal sex of the individual. This mode of inheritance is in contrast to the inheritance of traits on autosomal chromosomes, where both sexes have the same probability of inheritance. Since humans have many more genes on the X than the Y, there are many more X-linked traits than Y-linked traits.

In mammals, the female is the homozygous sex, with two X chromosomes (XX), while the male is heterozygous, with one X and one Y chromosome (XY). Genes on the X or Y chromosome are called sex linked genes.

In birds, the opposite is true: the male is the homozygous sex, having two Z chromosomes (ZZ), and the female (hen) is heterozygous, having one Z and one W chromosome (ZW).

Linkage Overview

X-linked recessive traits are expressed in all heterogametics, but are only expressed in those homogametics that are homozygous for the recessive allele. For example, an X-linked recessive allele in humans causes hemophilia. Hemophilia is much more common in males than females because males are hemizygous - they only have one copy of the gene in question - and therefore express the trait when they inherit one mutant allele.

In contrast, a female must inherit two mutant alleles, a less frequent event since the mutant allele is rare in the population. Tsarevich Alexei of Russia was the most famous sufferer of X-linked hemophilia, and his disease may have played an important role in the overthrow of the imperial regime, which changed the course of history for millions of people.

The incidence of recessive X-linked phenotypes in females is the square of that in males (squaring a proportion less than one gives an outcome closer to 0 than the original). If 1 in 20 males in a human population are green colour blind, then 1 in 400 females in the population are expected to be colour blind $(/_{20})*(/_{20})$.

X-linked traits are maternally inherited from carrier mothers or from an affected father. Each son born to a carrier mother has a 50% probability of inheriting the X-chromosome carrying the mutant allele. There are a few Y-linked traits; these are inherited from the father.

In classical genetics, a reciprocal cross is performed to test if a trait is sex-linked.

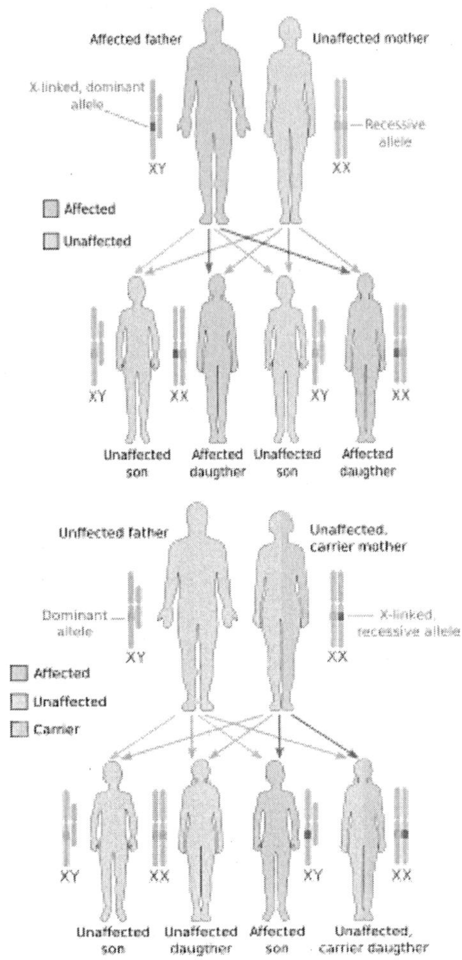

Illustration of some X-linked heredity outcomes (A) the affected father has one X-linked dominant allele, the mother is homozygous for the recessive allele: only daughters (all) will be affected. (B) the affected mother is hemizygous with one copy of the X-linked dominant allele: both daughters and sons will have 50% probability to be affected. C) the hemizygous mother is called "carrier" because she has one copy of the recessive allele: sons will have 50% probability to be affected, 50% of unaffected daughters will become carriers like their mother.

Autosomal Recessive Inheritance

Characteristics of Autosomal Recessive Inheritance: • An individual will be a “carrier” if they posses one mutated allele and one normal gene copy. • There is a 50% chance that a carrier will transmit a mutated gene to a child. • If two carrier parents have a child there is a: o 25% chance that both will transmit the mutated gene; in this case, the child will inherit only mutated copies of the gene from both the mother and the father and thus will be affected with the disorder. o 50% chance that one carrier parent will transmit the mutated gene and the other will transmit the normal gene; in this case, the child will have one mutated gene and one normal gene and will be a carrier of the disorder. o 25% chance that both carrier parents will transmit the normal gene; in this case the child will have only normal genes and will not be affected and will not be a carrier.

Autosomal Dominant Inheritance

Characteristics of Autosomal Dominant Inheritance: • The child of an affected parent has a 50% chance of inheriting the parent's mutated allele and thus being affected with the disorder. • A mutation can be transmitted by either the mother or the father. All children, regardless of gender, have an equal chance of inheriting the mutation

Examples: • Tuberous sclerosis • Marfan syndrome • Neurofibromatosis • Huntington's disease • Retinoblastoma • Waardenburg syndrome • Myotonic dystrophy • Familial hypercholestrolemia (LDL receptor defect Type IIa) • Adult polycystic kidney disease • von Hippel Lindau • Familial adenomatous polyposis and Peutz Jeghers Syndrome • Hereditory spherocytosis • Achondroplasia • Ehlor's Danlos (vascular type) • Acute intermittent porphyria • Hypertrophic Obstructive Cardiomyopathy (HOCM) • Von Willebrand Disease • Polydactyly • Osteogenesis Imperfecta (Except Type VII) • Hereditary hemorrhagic telengiactasia (Osler-weber-rendu syndrome) • Osteopetrosis Type II (Adult type) • Hypokalemic Periodic Paralysis

Examples:

• cystic fibrosis, • sickle-cell disease, • Tay-Sachs disease, • Niemann-Pick disease, • spinal muscular atrophy, • Roberts's syndrome.

X-Linked Dominant Inheritance

Characteristics of X-Linked Dominant Inheritance: • A male or female child of an affected mother has a 50% chance of inheriting the mutation and thus being affected with the disorder. • All female children of an affected father will be affected (daughters possess their fathers' X-chromosome). • No male children of an affected father will be affected (sons do not inherit their fathers' X-chromosome). Examples: • Alport's syndrome • Aarsog's syndrome • Coffin - Lowry syndrome (CLS) • idiopathic hypoparathyroidism • incontinentia pigmenti • ornithine carbamoyl transferase deficiency • Rett syndrome (RS) • vitamin D resistant rickets • fragile X syndrome

X-Linked Recessive Inheritance

Characteristics of X-Linked Recessive Inheritance: • Females possessing one X-linked recessive mutation are considered carriers and will generally not manifest clinical symptoms of the disorder. • All males possessing an X-linked recessive mutation will be affected (males have a single X-chromosome and therefore have only one copy of X-linked genes). • All offspring of a carrier female have a 50% chance of inheriting the mutation. • All female children of an affected father will be carriers (daughters posses their fathers' X-chromosome). • No male children of an affected father will be affected (sons do not inherit their fathers' X-chromosome).

Examples:

• Lesch-Nyhan Syndrome • Duchene Muscular Dystrophy • Hunter's Disease • Menkes Disease (Kinky hair syndrome) • Glucose 6 Phosphate Dehydrogenase Deficiency • Hemophilia A and B • Fabry's Disease • Wiskott-Aldrich Syndrome • Bruton's Aggamaglobulinemia • Color Blindness • Complete Androgen Insensitivity • Congenital Aqueductal stenosis (hydrocephalus) • Inherited Nephrogenic Diabetes Insipidus

Y-linked

- Various failures in the SRY genes

Sex-linked Traits in Other aAnimals

- Calico or tortoiseshell cats, which have orange-and-black fur
- White eyes in Drosophila melanogaster flies—the first sex-linked gene ever discovered
- Evolution of dominance

The evolution of dominance concerns the evolution of genetic dominance. The central argument, that modifier genes act upon other genes to make them dominant or recessive, and that these are then themselves subject to natural selection was first proposed by the British population geneticist Ronald Fisher in 1928, and expanded upon in his book The Genetical Theory of Natural Selection. However,

Sewall Wright and J.B.S. Haldane believed that the main explanation for dominance should be based on physiological factors, and that selection for modifiers was not a primary force. This led to a major fallout between Fisher and Wright. Subsequent works, particularly in molecular biology and biochemistry, have tended to favour Wright's view without completely excluding that the Fisherian argument may, in some circumstances, apply.

Traits Previously Believed to be Mendelian

Some traits were previously believed to be Mendelian, but their inheritance is (probably) based on more complex genetic models , possibly involving more than one gene. These include :

- Eye color
- Hair color
- Morton's toe
- Tongue rolling
- Widow's peak (allele)
- Hitchhiker's thumb (recessive)
- Cleft chin (dominant)
- Dettached (dominant) or attached (recessive) earlobes

Autosomal Recessive

Two copies of the gene must be mutated for a person to be affected by an autosomal recessive disorder. An affected person usually has unaffected parents who each carry a single copy of the mutated gene (and are referred to as carriers). Two unaffected people who each carry one copy of the mutated gene have a 25% chance with each pregnancy of having a child affected by the disorder. Examples of this type of disorder are cystic fibrosis, sickle-cell disease, Tay-Sachs disease, Niemann-Pick disease, spinal muscular atrophy, and Roberts syndrome. Certain other phenotypes, such as wet versus dry earwax, are also determined in an autosomal recessive fashion.

X-linked dominant

X-linked dominant inheritance, sometimes referred to as X-linked dominance, is a mode of genetic inheritance by which a dominant gene is carried on the X chromosome. As an inheritance pattern, it is less common than the X-linked recessive type. In medicine, X-linked dominant inheritance indicates that a gene responsible for a genetic disorder is located on the X chromosome, and only one copy of the allele is sufficient to cause the disorder when inherited from a parent who has the disorder. X-linked dominant traits do not necessarily affect males more than females (unlike X-linked recessive traits). The exact pattern of inheritance varies, depending on whether the father or the mother has the trait of interest. All daughters of an affected father will also be affected but none of his sons will be affected (unless the mother is also affected). In addition, the mother of an affected son is also affected (but not necessarily the other way round).

X-linked dominant disorders are caused by mutations in genes on the X chromosome. Only a few disorders have this inheritance pattern, with a prime example being X-linked hypophosphatemic rickets. Males and females are both affected in these disorders, with males typically being more severely affected than females. Some X-linked dominant conditions such as Rett syndrome, incontinentia pigmenti type 2 and Aicardi syndrome are usually fatal in males either in utero or shortly after birth, and are therefore predominantly seen in females. Exceptions to this finding are extremely rare cases in which boys with Klinefelter syndrome (47,XXY) also inherit an X-linked dominant condition and exhibit symptoms more similar to those of a female in terms of disease severity.

The chance of passing on an X-linked dominant disorder differs between men and women. The sons of a man with an X-linked dominant disorder will all be unaffected (since they receive their father's Y chromosome), and his daughters will all inherit the condition.

A woman with an X-linked dominant disorder has a 50% chance of having an affected fetus with each pregnancy, although it should be noted that in cases such as incontinentia pigmenti only female offspring are generally viable. In addition, although these conditions do not alter fertility per se, individuals with Rett syndrome or Aicardi syndrome rarely reproduce. Genetics

As the X chromosome is one of the sex chromosomes (the other being the Y chromosome), X-linked inheritance is determined by the gender of the parent carrying a specific gene and can often seem complex. This is due to the fact that, typically, females have two copies of the X-chromosome, while males have only one copy. The difference between dominant and recessive inheritance patterns also plays a role in determining the chances of a child inheriting an X-linked disorder from their parentage. Males are normally hemizygous for the X chromosome, having only one copy. As a result, X-linked dominant disorders usually show higher expressivity in males than females.

Inheritance

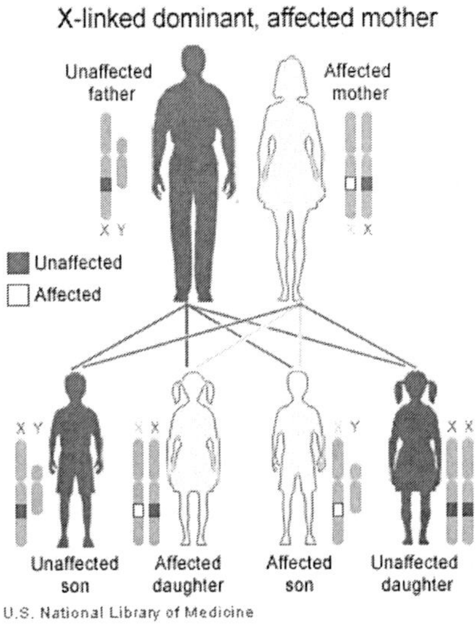

Figure:

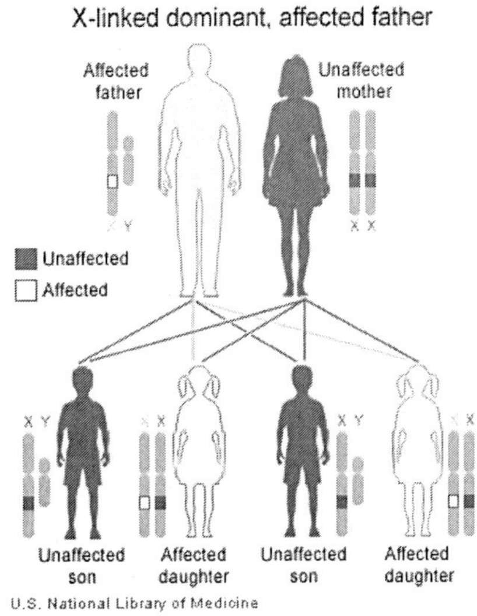

Figure:

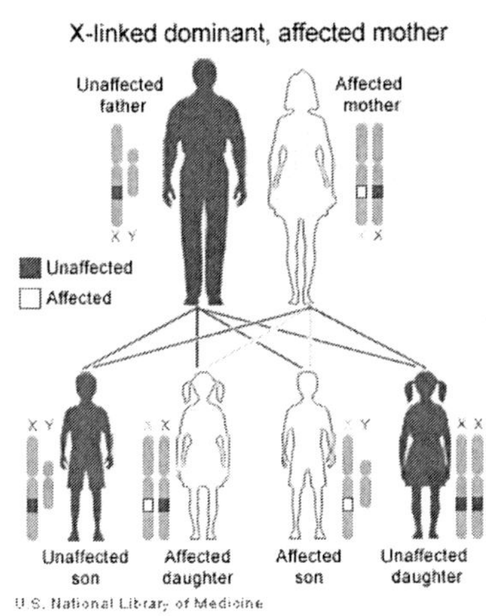

Figure:

X-linked dominant inheritance works differently depending upon whether the mother (left image) or father (right image) is the carrier of a gene that causes a disease or disorder. In X-linked dominant inheritance, when the mother alone is the carrier of a mutated, or defective gene associated with a disease or disorder; she herself will have the disorder. Her children will inherit the disorder as follows:

- Of her daughters and sons: 50% will have the disorder, 50% will be completely unaffected. Children of both genders have an even chance of receiving either of their mother's two X chromosomes, one of which contains the defective gene in question.

When the father alone is the carrier of a defective gene associated with a disease or disorder, he too will have the disorder. His children will inherit the disorder as follows:

- Of his daughters: 100% will have the disorder, since all of his daughters will receive one copy of his single X chromosome.
- Of his sons: none will have the disorder; sons do not receive an X chromosome from their father.

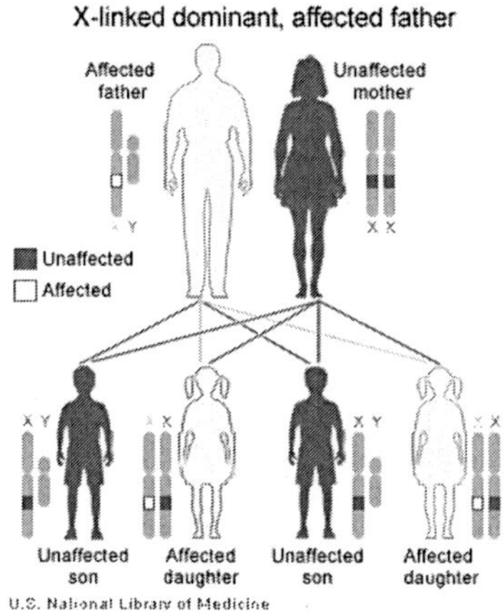

If both parents were carriers of a defective gene associated with a disease or disorder, they would both have the disorder. Their children would inherit the disorder as follows:

- Of their daughters: 100% will have the disorder, since all of the daughters will receive a copy of their father's X chromosome.
- Of the sons: 50% will have the disorder, 50% will be completely unaffected. Sons have an equal chance of receiving either of their mother's X chromosomes.

In such a case, where both parents carry and thus are affected by an X-linked dominant disorder, the chance of a daughter receiving two copies of the X chromosome with the defective gene is 50%, since daughters receive one copy of the X chromosome from both parents. Were this to occur with an X-linked dominant disorder, that daughter would likely experience a more severe form. Some X-linked dominant conditions such as Aicardi Syndrome are fatal to boys, therefore only girls with these conditions survive. Similarly, individuals with Klinefelter's Syndrome are referred to as "47,XXY Males".

X-linked Hypophosphatemia

X-linked hypophosphatemia (XLH), also called X-linked dominant hypophosphatemic rickets, X-linked vitamin d-resistant rickets or hypophosphatemic vitamin d-resistant rickets (HPDR), is an X-linked dominant form of rickets (or osteomalacia) that differs from most cases of rickets in that ingestion of vitamin D is relatively ineffective. It can cause bone deformity including short stature and genu varum (bow leggedness). It is associated with a mutation in the PHEX gene sequence (Xp.22) and subsequent inactivity of the PHEX protein. The prevalence of the disease is 1:20000. The leg deformity can be treated with Ilizarov frames and CHAOS surgery.

Cause and Genetics

XLH is associated with a mutation in the PHEX gene sequence, located on the human X chromosome at location Xp22.2-p22.1. The mutation results in altered (or missing) activity of the PHEX protein, which inactivates hormone-like substances (phosphatonins) that promote phosphate excretion. The resulting excess excretion of phosphate impairs bone mineralization. Biochemically, XLH is recognized by hypophosphatemia and inappropriately low level of calcitriol (1,25-$(OH)_2$ vitamin D_3).

The disorder is inherited in an X-linked dominant manner. This means the defective gene responsible for the disorder (PHEX) is located on the X chromosome, and only one copy of the defective gene is sufficient to cause the disorder when inherited from a parent who has the disorder. Males are normally hemizygous for the X chromosome, having only one copy. As a result, X-linked dominant disorders usually show higher expressivity in males than females.

As the X chromosome is one of the sex chromosomes (the other being the Y chromosome), X-linked inheritance is determined by the gender of the parent carrying a specific gene and can often seem complex. This is because, typically, females have two copies of the X-chromosome and males have only one copy. The difference between dominant and recessive inheritance patterns also plays a role in determining the chances of a child inheriting an X-linked disorder from their parentage.

X-linked Recessive

X-linked recessive conditions are also caused by mutations in genes on the X chromosome. Males are more frequently affected than females, and the chance of passing on the disorder differs between men and women. The sons of a man with an X-linked recessive disorder will not be affected, and his daughters will carry one copy of the mutated gene. A woman who is a carrier of an X-linked recessive disorder (XX) has a 50% chance of having sons who are affected and a 50% chance of having daughters who carry one copy of the mutated gene and are therefore carriers.

X-linked recessive conditions include the serious diseases Hemophilia A, Duchenne muscular dystrophy, and Lesch-Nyhan syndrome as well as common and less serious conditions such as male pattern baldness and red-green color blindness.

X-linked recessive conditions can sometimes manifest in females due to skewed X-inactivation or monosomy X (Turner syndrome).

Sex Differences In Phenotype/Genotypes and Frequency

In humans, generally "men are affected and women are carriers" for two reasons.

The first is the simple statistical fact that if the X-chromosomes is a population that carry a particular X-linked mutation at a frequency of 'f' (for example, 1%) then that will be the frequency that men are likely to express the mutation (since they have only one X), while women will express it at a frequency of f(for example 1% * 1% = 0.01%) since they have two X's and hence two chances to get the normal allele. Thus, X-linked mutations tend to be rare in women.

Most Common

The most common X-linked recessive disorders are:

- Color blindness; a classic example of an X-linked trait because it is easy to explain the phenotype and it's relatively common 7% to 10% of men are red-green colorblind, making the above calculations predict 0.49% to 1% of women are carriers. Its commonness may be attributable to its not being a serious disability in most cases and an actual advantage in some situations (for example, not being distracted by some of the color in color based camouflage).It is also known as daltonism.
- Hemophilia A; another famous example because it was found in European royal families who intermarried and were famous enough that their pedigrees could be established and offered in textbooks as a "famous example" of an X-linked trait that had been documented in history books before mendelian genetics was understood.

- Duchenne muscular dystrophy; muscular dystrophy associated with mutations in the dystrophin gene, characterized by rapid progression of muscle degeneration, eventually leading to loss in ambulation, respiratory failure and death.
- Becker's muscular dystrophy; milder form of Duchenne, causes slowly progressive muscle weakness of the legs and pelvis.
- Hemophilia B; a blood clotting disorder caused by a mutation of the Factor IX gene, leading to a deficiency of Factor IX. It is rarer than haemophilia A. It's also called Christmas disease
- X-linked ichthyosis; a skin condition (ichthyosis) caused by the hereditary deficiency of the steroid sulfatase (STS) enzyme that affects 1 in 2000 to 1 in 6000 males.
- X-linked agammaglobulinemia (XLA); affects the body's ability to fight infection (origin of the name: A=no, gammaglobulin = Antibody). XLA patients do not generate mature B cells. B cells are part of the immune system and normally manufacture antibodies (also called immunoglobulins) which defends the body from infections (the humoral response). Patients with untreated XLA are prone to develop serious and even fatal infections.
- Glucose-6-phosphate dehydrogenase deficiency; may exhibit nonimmune hemolytic anemia in response to a number of causes, most commonly infection or exposure to certain medications or chemicals.

Less Common Disorders

This section does not cite any references or sources. Please help improve this section by adding citations to reliable sources. Unsourced material may be challenged and removed. (March 2011)

Theoretically, a mutation in any of the genes in the Category:Genes on chromosome.

X may cause disease, but below are some notable ones, with short description of symptoms:

- Adrenoleukodystrophy; leads to progressive brain damage, failure of the adrenal glands and eventually death.
- Alport syndrome; glomerulonephritis, endstage kidney disease, and hearing loss.
- Androgen insensitivity syndrome; variable degrees of undervirilization and/or infertility in XY persons of either gender

- Barth syndrome; metabolism distortion, delayed motor skills, stamina deficiency, hypotonia, chronic fatigue, delayed growth, cardiomyopathy, and compromised immune system.
- Centronuclear myopathy; where cell nuclei are abnormally located in skeletal muscle cells. In CNM the nuclei are located at a position in the center of the cell, instead of their normal location at the periphery.
- Charcot-Marie-Tooth disease (CMTX2-3); disorder of nerves (neuropathy) that is characterized by loss of muscle tissue and touch sensation, predominantly in the feet and legs but also in the hands and arms in the advanced stages of disease.
- Coffin-Lowry syndrome; severe mental retardation sometimes associated with abnormalities of growth, cardiac abnormalities, kyphoscoliosis as well as auditory and visual abnormalities.
- Fabry disease; A lysosomal storage disease causing anhidrosis, fatigue, angiokeratomas, burning extremity pain and ocular involvement.
- Hunter's Syndrome; potentially causing hearing loss, thickening of the heart valves leading to a decline in cardiac function, obstructive airway disease, sleep apnea, and enlargement of the liver and spleen.
- Hypohidrotic ectodermal dysplasia, presenting with hypohidrosis, hypotrichosis, hypodontia
- Kabuki syndrome; multiple congenital anomalies and mental retardation.
- Kennedy disease; muscle cramps and progressive weakness
- Lesch-Nyhan syndrome; neurologic dysfunction, cognitive and behavioral disturbances including self-mutilation, and uric acid overproduction (hyperuricemia)
- Lowe Syndrome; hydrophthalmia, cataracts, intellectual disabilities, aminoaciduria, reduced renal ammonia production and vitamin D-resistant rickets
- Menkes disease; sparse and coarse hair, growth failure, and deterioration of the nervous system
- Nasodigitoacoustic syndrome; mishaped nose, brachydactyly of the distal phalanges, sensorineural deafness
- Nonsyndromic deafness and X-linked nonsyndromic deafness; hearing loss

- Norrie disease; cataracts, leukocoria along with other developmental issues in the eye
- Occipital horn syndrome; deformations in the skeleton
- Ornithine transcarbamylase deficiency; developmental delay and mental retardation. Progressive liver damage, skin lesions, and brittle hair may also be seen
- Siderius X-linked mental retardation syndrome; cleft lip and palate with mental retardation and facial dysmorphism, caused by mutations in the histone demethylase PHF8
- Simpson-Golabi-Behmel syndrome; coarse faces with protruding jaw and tongue, widened nasal bridge, and upturned nasal tip
- Spinal muscular atrophy caused by UBE1 gene mutation; weakness due to loss of the motor neurons of the spinal cord and brainstem
- Wiskott-Aldrich syndrome; eczema, thrombocytopenia, immune deficiency, and bloody diarrhea
- X-linked Severe Combined Immunodeficiency (SCID); infections, usually causing death in the first years of life
- X-linked sideroblastic anemia; skin paleness, fatigue, dizziness and enlarged spleen and liver.

Y linkage

In mammals, Y-linkage refers to when a phenotypic trait is determined by an allele (or gene) on the Y chromosome. It is also known as holandric inheritance. The Y-chromosome is small and does not contain many genes, therefore few traits are Y-linked, and Y-linked diseases are rare. Because the only humans which have a Y chromosome are males, the genes are simply passed from father to son, with no interchromosomal genetic recombination.

Chromosome Y deletions are a frequent genetic cause of male infertility. Another example in humans of a y-linked trait was thought to be hairy ears (it may also be sex-limited). However, this has been discredited.

Y-linked

Y-linked disorders are caused by mutations on the Y chromosome. Because males inherit a Y chromosome from their fathers, every son of an affected father will be affected. Because females inherit an X chromosome from their fathers, female offspring of affected fathers are never affected.

Since the Y chromosome is relatively small and contains very few genes, there are relatively few Y-linked disorders. Often the symptoms include infertility, which may be circumvented with the help of some fertility treatments. Examples are male infertility and hypertrichosis pinnae.

Mitochondrial

Mitochondrial diseases are a group of disorders caused by dysfunctional mitochondria, the organelles that are the "powerhouses" of the cell. Mitochondria are found in every cell of the human body except red blood cells. Mitochondria convert the energy of food molecules into the ATP that powers most cell functions. Mitochondrial diseases are often caused by genetics or mutations to the mitochondrial DNA that affect mitochondria function. Mitochondrial diseases take on unique characteristics both because of the way the diseases are often inherited and because mitochondria are so critical to cell function.

The subclass of these diseases that have neuromuscular disease symptoms are often called a mitochondrial myopathy. There is some evidence for mitochondrial dysfunction as the molecular basis of bipolar disorder. In addition, classical mitochondrial diseases occur in a subset of individuals with autism and are usually caused by genetic anomalies or mitochondrial respiratory pathway deficits.

This type of inheritance, also known as maternal inheritance, applies to genes in mitochondrial DNA. Because only egg cells contribute mitochondria to the developing embryo, only mothers can pass on mitochondrial conditions to their children. An example of this type of disorder is Leber's hereditary optic neuropathy.

Classification

In addition to the mitochondrial myopathies, other examples include:

- Diabetes mellitus and deafness (DAD)
    - — This combination at an early age can be due to mitochondrial disease
    - — Diabetes mellitus and deafness can also be found together for other reasons
- Leber's hereditary optic neuropathy (LHON)
    - — Visual loss beginning in young adulthood
    - — Eye disorder characterized by progressive loss of central vision due to degeneration of the optic nerves and retina

- Wolff-Parkinson-White syndrome
- Multiple sclerosis-type disease
- Affects 1 in 50,000 people in Finland

- Leigh syndrome, subacute sclerosing encephalopathy
  - After normal development the disease usually begins late in the first year of life, although onset may occur in adulthood
  - A rapid decline in function occurs and is marked by seizures, altered states of consciousness, dementia, ventilatory failure
- Neuropathy, ataxia, retinitis pigmentosa, and ptosis (NARP)
  - Progressive symptoms as described in the acronym
  - Dementia
- Myoneurogenic gastrointestinal encephalopathy (MNGIE)
  - Gastrointestinal pseudo-obstruction
  - Neuropathy
- Myoclonic Epilepsy with Ragged Red Fibers (MERRF)
  - Progressive myoclonic epilepsy
  - "Ragged Red Fibers" – clumps of diseased mitochondria accumulate in the subsarcolemmal region of the muscle fiber and appear as "Ragged Red Fibers" when muscle is stained with modified Gömöri trichrome stain
  - Short stature
  - Hearing loss
  - Lactic acidosis
  - Exercise intolerance
- Mitochondrial myopathy, encephalomyopathy, lactic acidosis, stroke-like symptoms (MELAS)
- MtDNA depletion
  - Mitochondrial neurogastrointestinal encephalomyopathy (MNGIE)

Nota bene: Conditions such as Friedreich's ataxia can affect the mitochondria, but are not associated with mitochondrial proteins.

Symptoms

Symptoms include poor growth, loss of muscle coordination, muscle weakness, visual problems, hearing problems, learning disabilities, heart disease, liver disease, kidney disease, gastrointestinal disorders, respiratory disorders, neurological problems, autonomic dysfunction, and dementia.

Sources and Storage of Cells

To limit the risks of transplanted stem cell rejection or of severe graft-versus-host disease in allogeneic HSCT, the donor should preferably have the same human leukocyte antigens (HLA) as the recipient. About 25 to 30 percent of allogeneic HSCT recipients have an HLA-identical sibling. Even so-called "perfect matches" may have mismatched minor alleles that contribute to graft-versus-host disease.

Bone Marrow

In the case of a bone marrow transplant, the HSC are removed from a large bone of the donor, typically the pelvis, through a large needle that reaches the center of the bone. The technique is referred to as a bone marrow harvest and is performed under general anesthesia.

Peripheral Blood Stem Cells

Peripheral stem cell transplantation is a method of replacing blood-forming cells destroyed by cancer treatment. Immature blood cells (stem cells) in the circulating blood that are similar to those in the bone marrow are given to the patient after treatment. This helps the bone marrow recover and continue producing healthy blood cells. Transplantation may be autologous (an individual's own blood cells saved earlier), allogeneic (blood cells donated by someone else), or syngeneic (blood cells donated by an identical twin). Also called peripheral stem cell support. Peripheral blood stem cells are now the most common source of stem cells for allogeneic HSCT. They are collected from the blood through a process known as apheresis. The donor's blood is withdrawn through a sterile needle in one arm and passed through a machine that removes white blood cells. The red blood cells are returned to the donor. The peripheral stem cell yield is boosted with daily subcutaneous injections of Granulocyte-colony stimulating factor, serving to mobilize stem cells from the donor's bone marrow into the peripheral circulation.

Umbilical Cord Blood

Umbilical cord blood is obtained when a mother donates her infant's umbilical cord and placenta after birth. Cord blood has a higher concentration of HSC than is normally found in adult blood. However, the small quantity of blood obtained from an umbilical cord (typically about 50 mL) makes it more suitable for transplantation into small children than into adults. Newer techniques using ex-vivo expansion of cord blood units or the use of two cord blood units from different donors are being explored to allow cord blood transplants to be used in adults.

Cord blood is typically harvested from the umbilical cord of a child being born after preimplantation genetic diagnosis (PGD) for human leucocyte antigen (HLA) matching in order to donate to an ill sibling requiring HSCT.

Amniotic Fluid

It is also possible to extract hematopoietic stem cells from amniotic fluid for both autologous or heterologous use at the time of childbirth.

Storage of HSC

Unlike other organs, bone marrow cells can be frozen (cryopreserved) for prolonged periods, without damaging too many cells. This is a necessity with autologous HSC because the cells must be harvested from the recipient months in advance of the transplant treatment. In the case of allogeneic transplants, fresh HSC are preferred, in order to avoid cell loss that might occur during the freezing and thawing process. Allogeneic cord blood is stored frozen at a cord blood bank because it is only obtainable at the time of childbirth. To cryopreserve HSC, a preservative, DMSO, must be added, and the cells must be cooled very slowly in a controlled-rate freezer to prevent osmotic cellular injury during ice crystal formation. HSC may be stored for years in a cryofreezer, which typically utilizes liquid nitrogen.

Conditioning Regimens

Myeloablative Transplants

The chemotherapy or irradiation given immediately prior to a transplant is called the conditioning or preparative regimen, the purpose of which is to help eradicate the patient's disease prior to the infusion of HSC and to suppress immune reactions. The bone marrow can be ablated with dose-levels that cause minimal injury to other tissues. In allogeneic transplants a combination of cyclophosphamide with busulfan or total body irradiation is commonly employed. This treatment also has an immunosuppressive effect that prevents rejection of the HSC by the recipient's immune system. The post-transplant prognosis often includes acute and chronic graft-versus-host disease that may be life-threatening. However in certain leukemias this can coincide with protection against cancer relapse owing to the graft versus tumor effect. Autologous transplants may also use similar conditioning regimens, but many other chemotherapy combinations can be used depending on the type of disease.

Non-myeloablative Allogeneic Transplants

This is a newer treatment approach using lower doses of chemotherapy and radiation, which are too low to eradicate all the bone marrow cells of a recipient. Instead, non-myeloablative transplants run lower risks of serious infections and transplant-related mortality while relying upon the graft versus tumor effect to resist the inherent increased risk of cancer relapse.

Also significantly, while requiring high doses of immunosuppressive agents in the early stages of treatment, these doses are less than for conventional transplants. This leads to a state of mixed chimerism early after transplant where both recipient and donor HSC coexist in the bone marrow space. Decreasing doses of immunosuppressive therapy then allows donor T-cells to eradicate the remaining recipient HSC and to induce the graft versus tumor effect. This effect is often accompanied by mild graft-versus-host disease, the appearance of which is often a surrogate for the emergence of the desirable graft versus tumor effect, and also serves as a signal to establish an appropriate dosage level for sustained treatment with low levels of immunosuppressive agents.

Because of their gentler conditioning regimens, these transplants are associated with a lower risk of transplant-related mortality and therefore allow patients who are considered too high-risk for conventional allogeneic HSCT to undergo potentially curative therapy for their disease. These new transplant strategies are still somewhat experimental, but are being used more widely on elderly patients unfit for myeloablative regimens and for whom the higher risk of cancer relapse may be acceptable.

Engraftment

After several weeks of growth in the bone marrow, expansion of HSC and their progeny is sufficient to normalize the blood cell counts and reinitiate the immune system. The offspring of donor-derived hematopoietic stem cells have been documented to populate many different organs of the recipient, including the heart, liver, and muscle, and these cells had been suggested to have the abilities of regenerating injured tissue in these organs. However, recent research has shown that such lineage infidelities does not occur as a normal phenomenon.

Complications

HSCT is associated with a high treatment-related mortality in the recipient (10% or higher), which limits its use to conditions that are themselves life-threatening. Major complications are veno-occlusive disease, mucositis, infections (sepsis), graft-versus-host disease and the development of new malignancies.

Infection

Bone marrow transplantation usually requires that the recipient's own bone marrow be destroyed ("myeloablation"). Prior to "engraftment" patients may go for several weeks without appreciable numbers of white blood cells to help fight infection.

This puts a patient at high risk of infections, sepsis and septic shock, despite prophylactic antibiotics. However, antiviral medications, such as acyclovir and valacyclovir, are quite effective in prevention of HSCT-related outbreak of herpetic infection in seropositive patients.

The immunosuppressive agents employed in allogeneic transplants for the prevention or treatment of graft-versus-host disease further increase the risk of opportunistic infection. Immunosuppressive drugs are given for a minimum of 6-months after a transplantation, or much longer if required for the treatment of graft-versus-host disease.

Transplant patients lose their acquired immunity, for example immunity to childhood diseases such as measles or polio. For this reason transplant patients must be re-vaccinated with childhood vaccines once they are off immunosuppressive medications.

Veno-occlusive Disease

Severe liver injury can result from hepatic veno-occlusive disease (VOD). Elevated levels of bilirubin, hepatomegaly and fluid retention are clinical hallmarks of this condition. There is now a greater appreciation of the generalized cellular injury and obstruction in hepatic vein sinuses, and hepatic VOD has lately been referred to as sinusoidal obstruction syndrome (SOS). Severe casesof SOS are associated with a high mortality rate. Anticoagulants or defibrotide may be effective in reducing the severity of VOD but may also increase bleeding complications. Ursodiol has been shown to help prevent VOD, presumably by facilitating the flow of bile.

Mucositis

The injury of the mucosal lining of the mouth and throat is a common regimen-related toxicity following ablative HSCT regimens.

It is usually not life-threatening but is very painful, and prevents eating and drinking. Mucositis is treated with pain medications plus intravenous infusions to prevent dehydration and malnutrition.

Graft-versus-host Disease

Graft-versus-host disease (GVHD) is a common complication after a stem cell transplant or bone marrow transplant from another person (an allogeneic transplant). Immune cells (white blood cells) in the donated marrow or stem cells (the graft) recognize the recipient (the host) as "foreign". The transplanted immune cells then attack the host's body cells. Graft versus host disease can also take place during a blood transfusion under certain circumstances.

Graft-versus-host disease (GVHD) is an inflammatory disease that is unique to allogeneic transplantation. It is an attack of the "new" bone marrow's immune cells against the recipient's tissues. This can occur even if the donor and recipient are HLA-identical because the immune system can still recognize other differences between their tissues. It is aptly named graft-versus-host disease because bone marrow transplantation is the only transplant procedure in which the transplanted cells must accept the body rather than the body accepting the new cells. Acute graft-versus-host disease typically occurs in the first 3 months after transplantation and may involve the skin, intestine, or the liver.. High-dose corticosteroids such as prednisone are a standard treatment; however this immuno-suppressive treatment often leads to deadly infections. Chronic graft-versus-host disease may also develop after allogeneic transplant. It is the major source of late treatment-related complications, although it less often results in death. In addition to inflammation, chronic graft-versus-host disease may lead to the development of fibrosis, or scar tissue, similar to scleroderma; it may cause functional disability and require prolonged immunosuppressive therapy. Graft-versus-host disease is usually mediated by T cells, which react to foreign peptides presented on the MHC of the host .

Causes

According to the 1966 Billingham Criteria, 3 criteria must be met in order for GVHD to occur.

- An immuno-competent graft is administered, with viable and functional immune cells.
- The recipient is immunologically disparate - histo-incompatible.
- The recipient is immuno-compromised and therefore cannot destroy or inactivate the transplanted cells.

After bone marrow transplantation, T cells present in the graft, either as contaminants or intentionally introduced into the host, attack the tissues of the transplant recipient after perceiving host tissues as antigenically foreign. The T cells produce an excess of cytokines, including TNF-á and interferon-gamma (IFNã). A wide range of host antigens can initiate graft-versus-host-disease, among them the human leukocyte antigens (HLAs). However, graft-versus-host disease can occur even when HLA-identical siblings are the donors. HLA-identical siblings or HLA-identical unrelated donors often have genetically different proteins (called minor histocompatibility antigens) that can be presented by Major histocompatibility complex (MHC) molecules to the donor's T-cells, which see these antigens as foreign and so mount an immune response.

While donor T-cells are undesirable as effector cells of graft-versus-host-disease, they are valuable for engraftment by preventing the recipient's residual immune system from rejecting the bone marrow graft (host-versus-graft). In addition, as bone marrow transplantation is frequently used to treat cancer, mainly leukemias, donor T-cells have proven to have a valuable graft-versus-tumor effect. A great deal of current research on allogeneic bone marrow transplantation involves attempts to separate the undesirable graft-vs-host-disease aspects of T-cell physiology from the desirable graft-versus-tumor effect.

Types

In the clinical setting, graft-versus-host-disease is divided into acute and chronic forms.

- The acute or fulminant form of the disease (aGVHD) is normally observed within the first 100 days post-transplant, and is a major challenge to transplants owing to associated morbidity and mortality.
- The chronic form of graft-versus-host-disease (cGVHD) normally occurs after 100 days. The appearance of moderate to severe cases of cGVHD adversely influences long-term survival.

This distinction is not arbitrary: Acute and chronic graft-versus-host-disease appear to involve different immune cell subsets, different cytokine profiles, somewhat different host targets, and respond differently to treatment.

Clinical Manifestation

In the classical sense, acute graft-versus-host-disease is characterized by selective damage to the liver, skin (rash), and mucosa, and the gastrointestinal tract. Newer research indicates that other graft-versus-host-disease target organs include the immune system (the hematopoietic system, e.g., the bone marrow and the thymus) itself, and the lungs in the form of idiopathic pneumonitis. Chronic graft-versus-host-disease also attacks the above organs, but over its long-term course can also cause damage to the connective tissue and exocrine glands.

Acute GVHD of the GI tract can result in severe intestinal inflammation, sloughing of the mucosal membrane, severe diarrhea, abdominal pain, nausea, and vomiting. This is typically diagnosed via intestinal biopsy. Liver GVHD is measured by the bilirubin level in acute patients. Skin GVHD results in a diffuse maculopapular rash, sometimes in a lacy pattern.

It appears in both acute and chronic GVHD, mucosal damage to the vagina can result in severe pain and scarring. This can result in an inability to have sexual intercourse. Acute GVHD is staged as follows: overall grade (skin-liver-gut) with each organ staged individually from a low of 1 to a high of 4. Patients with grade IV GVHD usually have a poor prognosis. If the GVHD is severe and requires intense immunosuppression involving steroids and additional agents to get under control, the patient may develop severe infections as a result of the immunosuppression and may die of infection.

In the oral cavity, chronic graft-versus-host-disease manifests as lichen planus with a higher risk of malignant transformation to oral squamous cell carcinoma in comparison to the classical oral lichen planus. Graft-versus-host-disease-associated oral cancer may have more aggressive behavior with poorer prognosis, when compared to oral cancer in non-hematopoietic stem cell transplantation patients.

Transfusion-associated GVHD

Transfusion-associated graft versus host disease (TA-GvHD) is a rare complication of blood transfusion, in which the donor T lymphocytes mount an immune response against the recipient's lymphoid tissue. Donor lymphocytes are usually identified as foreign and destroyed by the recipient's immune system.

However, in situations where the recipient is immunocompromised (inborn immunodeficiency, acquired immunodeficiency, malignancy), or when the donor is homozygous and the recipient is heterozygous for an HLA haplotype (as can occur in directed donations from first-degree relatives), the recipient's immune system is not able to destroy the donor lymphocytes. This can result in graft versus host disease.

This type of GVHD is associated with transfusion of un-irradiated blood to immunocompromised recipients. It can also occur in situations in which the blood donor is homozygous and the recipient is heterozygous for an HLA haplotype. It is associated with higher mortality (80-90%) due to involvement of bone marrow lymphoid tissue, however the clinical manifestations are similar to GVHD resulting from bone marrow transplantation. Transfusion-associated GVHD is rare in modern medicine. It is almost entirely preventable by controlled irradiation of blood products to inactivate the white blood cells (including lymphocytes) within. Epidemiology and pathogenesis

The incidence in immunocompromised patients receiving blood transfusions is estimated to be 0.1 - 1.0%, mortality around 80 - 90%. Mortality is higher in TA-GvHD than in GvHD associated with bone marrow transplantation, where the engrafted lymphoid cells in the bone marrow are of donor origin; therefore, the immune reaction is not directed against them. The most common cause of death in TA-GvHD is infections and hemorrhages, secondary to pancytopenia and liver dysfunction.

Presentation and Diagnosis

Clinical Manifestations

The clinical presentation is the same as GvHD occurring in other settings, such as bone marrow transplantation. TA-GvHD can develop four to thirty days after the transfusion. Typical symptoms include:

- Fever
- Erythematous maculopapular rash, which can progress to generalised erythroderma
- Toxic epidermal necrolysis in extreme cases

Other symptoms can include cough, abdominal pain, vomiting, and profuse diarrhea (up to 8 liters/day).

Laboratory Manifestations

Laboratory findings include pancytopenia, abnormal liver enzymes, and electrolyte imbalance (when diarrhea is present).

Diagnosis

TA-GvHD can be suspected from a biopsy of the affected skin, and established by HLA analysis of the circulating lymphocytes. This testing can identify circulating lymphocytes with a different HLA type than the tissue cells of the host.

Treatment and Prevention

Treatment is only supportive, as no available form of therapy has proven effective in treating TA-GvHD.

Prevention includes gamma irradiation of the lymphocyte-containing blood products. This procedure should be performed in transfusions when:

- The recipient is immunocompromised
- The blood components are from a family donor
- HLA-matched platelets are transferred.

Another means of prevention is the use of third- or fourth-generation leukoreduction filters, although the efficacy of this procedure has not yet been documented.

In Thymus Transplantation

Thymus transplantation may be said to be able to cause a special type of GVHD because the recipients thymocytes would use the donor thymus cells as models when going through the negative selection to recognize self-antigens, and could therefore still mistake own structures in the rest of the body for being non-self. This is a rather indirect GVHD because it is not directly cells in the graft itself that causes it but cells in the graft that make the recipient's T cells act like donor T cells. It can be seen as a multiple-organ autoimmunity in xenotransplantation experiments of the thymus between different species. Autoimmune disease is a frequent complication after human allogeneic thymus transplantation, found in 42% of subjects over 1 year post transplantation. However, this is partially explained by the fact that the indication itself, that is, complete DiGeorge syndrome, increases the risk of autoimmune disease.

Prevention

- DNA-based tissue typing allows for more precise HLA matching between donors and transplant patients, which has been proven to reduce the incidence and severity of GVHD and to increase long-term survival.
- The T-cells of umbilical cord blood (UCB) have an inherent immunological immaturity, and the use of UCB stem cells in unrelated donor transplants has a reduced incidence and severity of GVHD.
- Methotrexate, ciclosporin and tacrolimus are common drugs used for GVHD prophylaxis.

- Graft-versus-host-disease can largely be avoided by performing a T-cell-depleted bone marrow transplant. However, these types of transplants come at a cost of diminished graft-versus-tumor effect, greater risk of engraftment failure, or cancer relapse, and general immunodeficiency, resulting in a patient more susceptible to viral, bacterial, and fungal infection. In a multi-center study, disease-free survival at 3 years was not different between T cell-depleted and T cell-replete transplants.

Treatment of GVHD

Intravenously administered corticosteroids, such as prednisone, are the standard of care in acute GVHD and chronic GVHD. The use of these corticosteroids is designed to suppress the T-cell-mediated immune onslaught on the host tissues; however, in high doses, this immune-suppression raises the risk of infections and cancer relapse. Therefore, it is desirable to taper off the post-transplant high-level steroid doses to lower levels, at which point the appearance of mild GVHD may be welcome, especially in HLA mis-matched patients, as it is typically associated with a graft-versus-tumor effect.

The use of irradiated blood components prevent GVHD. The irradiation of blood eliminates the proliferative capacity of lymphocytes present in red cell, platelet and freshly collected plasma components. The penetrated photons of radiation beam into blood components cause the formation of electrically charged particles or secondary electrons. These electrons damage the DNA of lymphocytes either by direct interaction or by reacting initicially with cell water to form free radicals. The damaged lymphocytes are unable to proliferate in the host and therefore cannot mediate transfusion-associated GVHD.

Investigational Therapies for Graft-versus-host Disease

There are a large number of clinical trials either ongoing or recently completed in the investigation of graft-versus-host disease treatment and prevention.

Graft-versus-tumor Effect

Graft-versus-tumor effect (GVT) or "graft versus leukemia" effect is the beneficial aspect of the Graft-versus-Host phenomenon. For example, HSCT patients with either acute and in particular chronic graft-versus-host disease after an allogeneic transplant tend to have a lower risk of cancer relapse. This is due to a therapeutic immune reaction of the grafted donor T lymphocytes against the diseased bone marrow of the recipient. This lower rate of relapse accounts for the increased success rate of allogeneic transplants, compared to transplants from identical twins, and indicates that allogeneic HSCT is a form of immunotherapy. GVT is the major benefit of transplants that do not employ the highest immuno-suppressive regimens.

Graft versus tumor is mainly beneficial in diseases with slow progress, e.g. chronic leukemia, low-grade lymphoma, and some cases multiple myeloma. However, it is less effective in rapidly growing acute leukemias. If cancer relapses after HSCT, another transplant can be performed, infusing the patient with a greater quantity of donor white blood cells.

Oral Carcinoma

Patients after HSCT are at a higher risk for oral carcinoma. Post-HSCT oral cancer may have more aggressive behavior with poorer prognosis, when compared to oral cancer in non-HSCT patients.

Prognosis

Prognosis in HSCT varies widely dependent upon disease type, stage, stem cell source, HLA-matched status (for allogeneic HCST) and conditioning regimen.

A transplant offers a chance for cure or long-term remission if the inherent complications of graft versus host disease, immuno-suppressive treatments and the spectrum of opportunistic infections can be survived. In recent years, survival rates have been gradually improving across almost all populations and sub-populations receiving transplants.

Mortality for allogeneic stem cell transplantation can be estimated using the prediction model created by Sorror et al., using the Hematopoietic Cell Transplantation-Specific Comorbidity Index (HCT-CI). The HCT-CI was derived and validated by investigators at the Fred Hutchinson Cancer Research Center (Seattle, WA). The HCT-CI modifies and adds to a well-validated comorbidity index, the Charlson Comorbidity Index (CCI) (Charlson et al.) The CCI was previously applied to patients undergoing allogeneic HCT but appears to provide less survival prediction and discrimination than the HCT-CI scoring system.

Risks to Donor

The risks of a complication depend on patient characteristics, health care providers and the apheresis procedure, and the colony-stimulating factor used (G-CSF, GM-CSF). G-CSF drugs include Filgrastim (Neupogen, Neulasta), and lenograstim (Graslopin).

Drug Risks

Filgrastim is typically dosed in the 10 microgram/kg level for 4–5 days during the harvesting of stem cells. The documented adverse effects of filgrastim include splenic rupture (indicated by left upper abdominal or shoulder pain, risk 1 in 40000), Adult respiratory distress syndrome (ARDS), alveolar hemorrage, and allergic reactions (usually expressed in first 30 minutes, risk 1 in 300). In addition, platelet and hemoglobin levels dip post-procedure, not returning to normal until one month.

The question of whether patients over 65 react the same as patients under 65 has not been sufficiently examined. Coagulation issues and inflammation of atherosclerotic plaques are known to occur as a result of G-CSF injection. G-CSF has also been described to induce genetic changes in mononuclear cells of normal donors. There is evidence that myelodysplasia (MDS) or acute myeloid leukaemia (AML) can be induced by GCSF in susceptible individuals.

Access Risks

Blood was drawn peripherally in a majority of patients, but a central line to jugular/subclavian/femoral veins may be used in 16% of women and 4% of men. Adverse reactions during apheresis were experienced in 20% of women and 8% of men, these adverse events primarily consisted of numbness/tingling, multiple line attempts, and nausea.

Clinical Observations

A study involving 2408 donors (18–60 years) indicated that bone pain (primarily back and hips) as a result of filgrastim treatment is observed in 80% of donors by day 4 post-injection. This pain responded to acetaminophen or ibuprofen in 65% of donors and was characterized as mild to moderate in 80% of donors and severe in 10%. Bone pain receded post-donation to 26% of patients 2 days post-donation, 6% of patients one week post-donation, and <2% 1 year post-donation. Donation is not recommended for those with a history of back pain. Other symptoms observed in more than 40% of donors include myalgia, headache, fatigue, and insomnia. These symptoms all returned to baseline 1 month post-donation, except for some cases of persistent fatigue in 3% of donors. . In one metastudy that incorporated data from 377 donors, 44% of patients reported having adverse side effects after peripheral blood HSCT. Side effects included pain prior to the collection procedure as a result of GCSF injections, post-procedural generalized skeletal pain, fatigue and reduced energy.

Severe Reactions

A study that surveyed 2408 donors found that serious adverse events (requiring prolonged hospitalization) occurred in 15 donors (at a rate of 0.6%), although none of these events were fatal. Donors were not observed to have higher than normal rates of cancer with up to 4–8 years of follow up. One study based on a survey of medical teams covered approximately 24,000 peripheral blood HSCT cases between 1993 and 2005, and found a serious cardiovascular adverse reaction rate of about 1 in 1500. This study reported a cardiovascular-related fatality risk within the first 30 days HSCT of about 2 in 10000. For this same group, severe cardiovascular events were observed with a rate of about 1 in 1500. The most common severe adverse reactions were pulmonary edema/deep vein thrombosis, splenic rupture, and myocardial infarction. Haematological malignancy induction was comparable to that observed in the general population with only 15 reported cases within 4 years. ..

Donor Registration and Recruitment

At the end of 2010, 14.9 million people had registered their willingness to be a bone marrow donor with one of the 64 registries from 45 countries participating in Bone Marrow Donors Worldwide. 12.2 million of these registered donors had been ABDR typed, allowing easy matching. A further 453,000 cord blood units had been received by one of 44 cord blood units from 26 countries participating. The highest total number of bone marrow donors registered were those from the USA (6.4 million), and the highest number per capita were those from Cyprus (10.6% of the population).

Experimental HIV treatment

In 2007, a team of doctors in Berlin, Germany, including Gero Hütter, performed a stem cell transplant for a leukaemia patient who was also HIV-positive. From 60 matching donors, they selected a [CCR5]-Ä32 homozygous individual with two genetic copies of a rare variant of a cell surface receptor. This genetic trait confers resistance to HIV infection by blocking attachment of HIV to the cell. Roughly one in 1000 people of European ancestry have this inherited mutation, but it is rarer in other populations. The transplant was repeated a year later after a relapse. Over three years after the initial transplant and despite discontinuing antiretroviral therapy, researchers cannot detect HIV in the transplant recipient's blood or in various biopsies. Levels of HIV-specific antibodies have also declined, leading to speculation that the patient may have been functionally cured of HIV. However, scientists emphasise that this is an unusual case. Potentially fatal transplant complications (the "Berlin patient" suffered from graft-versus-host disease and leukoencephalopathy) mean that the procedure could not be performed in others with HIV, even if sufficient numbers of suitable donors were found.

Potential Treatments

Brain Damage

Stroke and traumatic brain injury lead to cell death, characterized by a loss of neurons and oligodendrocytes within the brain. Healthy adult brains contain neural stem cells which divide to maintain general stem cell numbers, or become progenitor cells. In healthy adult animals, progenitor cells migrate within the brain and function primarily to maintain neuron populations for olfaction (the sense of smell). Interestingly, in pregnancy and after injury, this system appears to be regulated by growth factors and can increase the rate at which new brain matter is formed. Although the reparative process appears to initiate following trauma to the brain, substantial recovery is rarely observed in adults, suggesting a lack of robustness. Stem cells may also be used to treat brain degeneration, such as in Parkinson's and Alzheimer's disease.

Cancer

The development of gene therapy strategies for treatment of intra-cranial tumours offers much promise, and has shown to be successful in the treatment of some dogs; although research in this area is still at an early stage. Using conventional techniques, brain cancer is difficult to treat because it spreads so rapidly. Researchers at the Harvard Medical School transplanted human neural stem cells into the brain of rodents that received intracranial tumours. Within days, the cells migrated into the cancerous area and produced cytosine deaminase, an enzyme that converts a non-toxic pro-drug into a chemotheraputic agent. As a result, the injected substance was able to reduce the tumor mass by 81 percent. The stem cells neither differentiated nor turned tumorigenic. Some researchers believe that the key to finding a cure for cancer is to inhibit proliferation of cancer stem cells. Accordingly, current cancer treatments are designed to kill cancer cells. However, conventional chemotherapy treatments cannot discriminate between cancerous cells and others. Stem cell therapies may serve as potential treatments for cancer. Research on treating

Lymphoma using adult stem cells is underway and has had human trials. Essentially, chemotherapy is used to completely destroy the patients own lymphocytes, and stem cells injected, eventually replacing the immune system of the patient with that of the healthy donor.

Spinal cord injury

A team of Korean researchers reported on November 25, 2003, that they had transplanted multipotent adult stem cells from umbilical cord blood to a patient suffering from a spinal cord injury and that following the procedure, she could walk on her own, without difficulty. The patient had not been able to stand up for roughly 19 years. For the unprecedented clinical test, the scientists isolated adult stem cells from umbilical cord blood and then injected them into the damaged part of the spinal cord.

According to the October 7, 2005 issue of The Week, University of California, Irvine researchers transplanted multipotent human fetal-derived neural stem cells into paralyzed mice, resulting in locomotor improvements four months later.

The observed recovery was associated with differentiation of transplanted cells into new neurons and oligodendrocytes- the latter of which forms the myelin sheath around axons of the central nervous system, thus insulating neural impulses and facilitating communication with the brain.

In January 2005, researchers at the University of Wisconsin–Madison differentiated human blastocyst stem cells into neural stem cells, then into pre-mature motor neurons, and finally into spinal motor neurons, the cell type that, in the human body, transmits messages from the brain to the spinal cord and subsequently mediates motor function in the periphery. The newly generated motor neurons exhibited electrical activity, the signature action of neurons. Lead researcher Su-Chun Zhang described the process as "[teaching] the blastocyst stem cells to change step by step, where each step has different conditions and a strict window of time."

Transformation of blastocyst stem cells into motor neurons had eluded researchers for decades. While Zhang's findings were a significant contribution to the field, the ability of transplanted neural cells to establish communication with neighboring cells remains unclear. Accordingly, studies using chicken embryos as a model organism can be an effective proof-of-concept experiment. If functional, the new cells could be used to treat diseases like Lou Gehrig's disease, muscular dystrophy, and spinal cord injuries.

Heart Damage

Several clinical trials targeting heart disease have shown that adult stem cell therapy is safe, effective, and equally efficient in treating old and recent infarcts. Adult stem cell therapy for treating heart disease was commercially available in at least five continents at the last count (2007).

Possible mechanisms of recovery include:

- Generation of heart muscle cells
- Stimulation of growth of new blood vessels to repopulate damaged heart tissue
- Secretion of growth factors
- Assistance via some other mechanism

It may be possible to have adult bone marrow cells differentiate into heart muscle cells.

Haematopoiesis (Blood Cell Formation)

The specificity of the human immune cell repertoire is what allows the human body to defend itself from rapidly adapting antigens. However, the immune system is vulnerable to degradation upon the pathogenesis of disease, and because of the critical role that it plays in overall defense, its degradation is often fatal to the organism as a whole.

Diseases of hematopoietic cells are called hematopathology. The specificity of the immune cells is what allows recognition of foreign antigens, causing further challenges in the treatment of immune disease. Identical matches between donor and recipient must be made for successful transplantation treatments, but matches are uncommon, even between first-degree relatives.

Research using both hematopoietic adult stem cells and embryonic stem cells has provided insight into the possible mechanisms and methods of treatment for many of these ailments.

Fully mature human red blood cells may be generated ex vivo by hematopoietic stem cells (HSCs), which are precursors of red blood cells. In this process, HSCs are grown together with stromal cells, creating an environment that mimics the conditions of bone marrow, the natural site of red blood cell growth. Erythropoietin, a growth factor, is added, coaxing the stem cells to complete terminal differentiation into red blood cells. Further research into this technique should have potential benefits to gene therapy, blood transfusion, and topical medicine.

Baldness

Hair follicles also contain stem cells, and some researchers predict research on these follicle stem cells may lead to successes in treating baldness through an activation of the stem cells progenitor cells. This treatment is expected to work by activating already existing stem cells on the scalp. Later treatments may be able to simply signal follicle stem cells to give off chemical signals to nearby follicle cells which have shrunk during the aging process, which in turn respond to these signals by regenerating and once again making healthy hair. Most recently, Dr. Aeron Potter of the University of California has claimed that stem cell therapy led to a significant and visible improvement in follicular hair growth . Results from his experiments are under review by the journal Science (journal).

Missing Teeth

In 2004, scientists at King's College London discovered a way to cultivate a complete tooth in mice and were able to grow them stand-alone in the laboratory. Researchers are confident that this technology can be used to grow live teeth in human patients. In theory, stem cells taken from the patient could be coaxed in the lab into turning into a tooth bud which, when implanted in the gums, will give rise to a new tooth, and would be expected to grow within two months. It will fuse with the jawbone and release chemicals that encourage nerves and blood vessels to connect with it. The process is similar to what happens when humans grow their original adult teeth. Many challenges remain, however, before stem cells could be a choice for the replacement of missing teeth in the future.

Bibliography

Ausubel, F.M. : *Current Protocols in Molecular Biology,* New York: John Wiley and Sons, 2009.

Ayala, F. J.: *The Genetic Structure of Populations*, W.H. Freeman & Co., San Francisco, California, 2007.

Bellairs, A.: *The Life of Reptiles*, Weidenfield and Nicholson, London, 2010.

Berg, Paul, and Singer, Maxine: *Dealing with Genes: The Language of Heredity. Mill Valley*, University Science Books, CA, 2012.

Bonnicksen, Andrea L.: *Crafting a Cloning Policy: From Dolly to Stem Cells*. Washington, DC: Georgetown University Press, 2013.

Bradley, W.G.: *Neurology in Clinical Practice*, Butterworth Heinemann, Boston, 2016.

Chambers, K. L.: *Biochemical Coevolution: Eugene,* Oregon State University Press, 2011.

Charles Hardin, Jennifer Edwards: *Cloning Gene Expression and Protein Purification*, Oxford University Press, Delhi, 2018.

Dawkins, R.: *The Selfish Gene*, Oxford University Press, New York, 2009.

Dhawan, Ruchi: *Cloning, Genetics and Stem Cell Technology : Development and Applications*, Axis Pub, Delhi, 2020.

Dobzhanshy, T.: *Genetics of the Evolutionary Process*: New York, Columbia University Press, 2012.

Edwards, Jennifer: *Cloning the Buddha : The Moral Impact of Biotechnology*, Health Harmony, Delhi, 2011.

Fisher, R. A.: *The Genetical Theory of Natural Selection*, Clarendon Press, Oxford, 2009.

Friedberg, E.C.: *DNA Repair*, New York, WH Freeman and Company, 2007.

Fukuyama F.: *Our Posthuman Future: Consequences of the Biotechnology Revolution,* Farrar, Straus and Giroux, NewYork, 2013.

Gall, Joseph G.: *Landmark Papers in Cell Biology,* Cold Spring Harbor Laboratory Press, Plainview, NY, 2012.

Garza-Valdes, L.A.: *The DNA of God?* Hodder and Stoughton, London, 2014.

Griffiths A.J.F.: *Introduction to Genetic Analysis,* W.H. Freeman and Company, New York, USA, 2016.

Halacy, D.S., Jr.: *Genetic Revolution, Shaping Life for Tomorrow,* Harper & Row, Publishers, New York, 2008.

Hall, J.E.: *Textbook of Medical Physiology*, Saunders, Philadelphia, 2016.

Hames, B. D.: *Genes and Embryos*, Oxford University Press, New York, 2006.

Harsh Bhaskar: *Basic Facts on Biochemistry*, Campus Books, Delhi, 2021.

Hartl D, Jones E: *Genetics: Analysis of Genes and Genomes,* Jones & Bartlett, 2016.

Heller C. :*Life: The Science of Biology*, Massachusetts: Sinauer Associates, 2018.

Jonoska, N.: *Self-Assembling DNA Graphs, DNA-Based Computers VIII*, Springer-Verlag, Berlin, 2014.

Kallenbach, N.R.: *Molecular Structure: Chemical Reactivity and Biological Activity,* Oxford University Press, Oxford, 2007.

Kass, Leon R.: *Human Cloning and Human Dignity: The Report of the President's Council on Bioethics*. New York: Public Affairs, 2013.

Kurzweil, Ray: *The Age of Spiritual Machines*, New York, Penguin Books, 2014.

Lauritzen, Paul: *Cloning and the Future of Human Embryo Research*. New York: Oxford University Press, 2011.

Lim, Hwa A.: *Multiplicity Yours: Cloning, Stem Cell Research and Regenerative Medicine*. Singapore: World Scientific, 2016.

Lindee, M.S. : *The DNA Mystique: The Gene as a Cultural Icon*. New York: W.H. Freeman, 2008.

Meffe, G.K. : *Principles of Conservation Biology,* Sunderland, MA: Sinauer Associates, 2016.

Nelkin, D. and M.S. Lindee: *The DNA Mystique: The Gene as a Cultural Icon*, W.H. Freeman, New York, 2008.

Pence, Gregory E.: *Cloning After Dolly: Who's Still Afraid?* Lanham, MD: Rowman & Littlefield, 2015.

Prakash, S. : *Cell Physiology*, Discovery, Delhi, 2020.

Ruse, Michael, and Aryne Sheppard: *Cloning: Responsible Science or Technomadness?* Amherst, NY: Prometheus, 2012.

Singh, Bimla: *Biochemistry and Genomic Revolution*, Vista International Pub, Delhi, 2017.

Stwertka, Eve and Albert: *Genetic Engineering*, Franklin Watts, New York, 2007.

Tripathi, Nirmal : *Molecular Biotechnology,* Crescent Publishing Corporation, Delhi, 2018.

Vasudev Bhatnagar: *Cell Science and Technology*, Campus Books, Delhi, 2019.

William, Kovacs D.: *An Introduction to Geotechnical Engineering*, Prentice Hall, UK, 2010.

Index

A

Algorithm, 3, 67, 69, 71, 80, 131, 153, 179, 180, 186.
Alternative Splicing Mechanisms, 48.
Analytical Technologies, 107.
Applied Memetic Engineering, 6.
Arbitrary Shapes, 87.

B

Biochemical mechanism, 44.
Bioinformatics, 67, 76, 77, 127, 133, 152, 186, 188, 189, 200, 213, 216, 230, 258.
Boolean Logic Assumption, 185.

C

Chain-termination Methods, 156.
Chemogenomics Approaches, 1.
Clinical Observations, 303.
Computational Methods, 77, 79, 111.
Computational Prediction of CRMs, 186.
Cultural Anthropology, 1.
Cytoplasmic Polyadenylation, 36, 37.

D

Development, 12, 13, 15, 16, 45, 66, 70, 73, 74, 81, 113, 119, 126, 129, 134, 137, 140, 143, 145, 147, 175, 201, 230, 255, 274, 293, 294, 295.

E

Evolution of Developmental Mechanisms, 212.
Experimental HIV treatment, 304.

F

Factors in Chemogenomics, 245.
Functional Nucleic acid Nanostructures, 88.
Future Methods, 233.
Future Research, 1.

G

General Organisation, 67.
Genome Assembly, 230.
Graft-Versus-Host Disease, 292, 293, 294, 295, 296, 297, 301, 305.

H

Health Benefits, 19.
Homology Modeling, 73, 80.
Human Behavioral Ecology, 2.

I

Imaging Techniques, 11, 12.

K

Key Cellular Functions, 107.
Knowledge Based Approaches, 169.

L

Language Acquisition, 7, 9, 12.
Limitations, 4, 105, 108, 122, 158, 250, 267.

M

Materials, 15, 86, 97, 98, 99, 141, 255.
Methodologies, 17, 126, 234.
Methods, 10, 19, 25, 41, 55, 59, 67, 68, 69, 71, 72, 73, 74, 75, 76, 77, 79, 80, 81, 83, 86, 88, 91, 92, 95, 98, 99, 101, 102, 104, 107, 109, 110, 111, 113, 114, 115, 120, 121, 122, 134, 135, 145, 152, 153, 154, 155, 157, 158, 160, 161, 167, 170, 186, 189, 198, 201, 202, 204, 205, 206, 207, 221, 222, 224, 233, 257, 258, 262, 263, 265, 268, 307.
Molecular Level Simulations, 215.

N

Nomenclature, 48, 190.
Nuclear Polyadenylation, 34.

O

Origin of Pseudogenes, 256.

P

Postulates, 100, 105.
Protein Databases, 127.
Protein Methods, 206.

Q

Quaternary Structure, 67, 77.

R

Recessive Genetic Diseases in Humans, 268.
Recruitment, 171, 304.
Relation to Proteome, 118.
Relationship to Docking Methods, 79.
Risks to Donor, 302.

S

Secondary Structure, 38, 67, 68, 69, 70, 72, 73, 76, 85, 171.
Separation Methods, 107.
Significance, 27, 50, 152.
Single gene Disorder, 236, 237.
Splicing Pathways, 41, 44.
Structural DNA Nanotechnology, 86.

T

Technology, 100, 101, 103, 104, 108, 110, 116, 118, 126, 145, 147, 154, 161, 162, 163, 166, 167, 197, 200, 221, 224, 225, 230, 231, 232, 234, 235, 259, 260, 261, 262, 266, 267, 268, 308.
Tertiary Structure, 68, 74, 81, 86, 121.
Transcription Factories, 26.

X

X-linked Hypophosphatemia, 285.

❑❑❑